"श्रीशाल"

The Power of Writing in Organizations

From Letters to Online Interactions

ORGANIZATION AND MANAGEMENT SERIES

Series Editors

Arthur P. Brief
University of Utah

Kimberly D. Elsbach
University of California, Davis

Michael Frese
University of Lueneburg and National University of Singapore

Ashforth (Au.): *Role Transitions in Organizational Life: An Identity-Based Perspective*
Bartel/Blader/Wrzesniewski (Eds.): *Identity and the Modern Organization*
Bartunek (Au.): *Organizational and Educational Change: The Life and Role of a Change Agent Group*
Beach (Ed.): *Image Theory: Theoretical and Empirical Foundations*
Brett/Drasgow (Eds.): *The Psychology of Work: Theoretically Based Empirical Research*
Brockner (Au.): *A Contemporary Look at Organizational Justice: Multiplying Insult Times Injury*
Chhokar/Brodbeck/House (Eds.): *Culture and Leadership Across the World: The GLOBE Book of In-Depth Studies of 25 Societies*
Darley/Messick/Tyler (Eds.): *Social Influences on Ethical Behavior in Organizations*
De Cremer/Tenbrunsel (Eds.): *Behavioral Business Ethics: Shaping an Emerging Field*
De Cremer/van Dick/Murnighan (Eds.): *Social Psychology and Organizations*
Denison (Ed.): *Managing Organizational Change in Transition Economies*
Dutton/Ragins (Eds.): *Exploring Positive Relationships at Work: Building a Theoretical and Research Foundation*
Elsbach (Au.): *Organizational Perception Management*
Earley/Gibson (Aus.): *Multinational Work Teams: A New Perspective*
Fayard/Metiu (Aus.): *The Power of Writing in Organizations: From Letters to Online Interactions*
Garud/Karnoe (Eds.): *Path Dependence and Creation*
Grandey/Diefendorff/Rupp (Eds.): *Emotional Labor in the 21st Century: Diverse Perspectives on Emotion Regulation at Work*
Harris (Ed.): *Handbook of Research in International Human Resource Management*
Jacoby (Au.): *Employing Bureaucracy: Managers, Unions, and the Transformation of Work in the 20th Century, Revised Edition*
Kossek/Lambert (Eds.): *Work and Life Integration: Organizational, Cultural and Individual Perspectives*

Kramer/Tenbrunsel/Bazerman (Eds.): *Social Decision Making: Social Dilemmas, Social Values and Ethical Judgments*

Lampel/Shamsie/Lant (Eds.): *The Business of Culture: Strategic Perspectives on Entertainment and Media*

Lant/Shapira (Eds.): *Organizational Cognition: Computation and Interpretation*

Lord/Brown (Aus.): *Leadership Processes and Follower Self-Identity*

Margolis/Walsh (Aus.): *People and Profits? The Search Between a Company's Social and Financial Performance*

Miceli/Dworkin/Near (Aus.): *Whistle-blowing in Organizations*

Nord/Connell (Aus.): *Rethinking the Knowledge Controversy in Organization Studies: A Generative Uncertainty Perspective*

Messick/Kramer (Eds.): *The Psychology of Leadership: Some New Approaches.*

Pearce (Au.): *Organization and Management in the Embrace of the Government*

Peterson/Mannix (Eds.): *Leading and Managing People in the Dynamic Organization*

Rafaeli/Pratt (Eds.): *Artifacts and Organizations: Beyond Mere Symbolism*

Riggio/Murphy/Pirozzolo (Eds.): *Multiple Intelligences and Leadership*

Roberts/Dutton (Eds.): *Exploring Positive Identities and Organizations: Building a Theoretical and Research Foundation*

Schneider/Smith (Eds.): *Personality and Organizations*

Smith (Ed.): *The People Make The Place: Dynamic Linkages Between Individuals and Organizations*

Thompson/Choi (Eds.): *Creativity and Innovation in Organizational Teams*

Thompson/Levine/Messick (Eds.): *Shared Cognition in Organizations: The Management of Knowledge*

Zaccaro/Marks/DeChurch (Eds.): *Multiteam Systems: An Organization Form for Dynamic and Complex Environments*

The Power of Writing in Organizations

From Letters to Online Interactions

Anne-Laure Fayard and
Anca Metiu

NEW YORK AND LONDON

First published 2013

by Routledge
711 Third Avenue, New York, NY 10017
Simultaneously published in the UK

by Routledge
27 Church Road, Hove, East Sussex BN3 2FA

Routledge is an imprint of the Taylor & Francis Group, an informa business
© 2013 Taylor & Francis

The right of Anne-Laure Fayard and Anca Metiu to be identified as authors of this work has been asserted by him/her in accordance with sections 77 and 78 of the Copyright, Designs and Patents Act 1988.

All rights reserved. No part of this book may be reprinted or reproduced or utilised in any form or by any electronic, mechanical, or other means, now known or hereafter invented, including photocopying and recording, or in any information storage or retrieval system, without permission in writing from the publishers.

Trademark notice: Product or corporate names may be trademarks or registered trademarks, and are used only for identification and explanation without intent to infringe.

Library of Congress Cataloging in Publication Data
Fayard, Anne-Laure.
The power of writing in organizations: From letters to online interactions / Anne-Laure Fayard, Anca Metiu. — 1st ed.
p. cm. — (Series in organization and management)
Includes bibliographical references and index.
ISBN 978-0-415-88256-9 (hardback : alk. paper)
1. Communication—Psychological aspects. 2. Business writing. 3. Communication in organizations. 4. Online social networks—Psychological aspects. 5. Communities—Social aspects. I. Metiu, Anca. II. Title.

HM1206.F393 2012
302—dc23

2012006309

ISBN: 978-0-415-88256-9 (hbk)
ISBN: 978-0-203-10273-2 (ebk)

Typeset in [Times]
by Apex CoVantage, LLC

Printed and bound in the United States of America by
Walsworth Publishing Company, Marceline, MO.

To Melchior, Jyoti, Guilhem, and Noa
To Alma and Jacques

Contents

List of Figures xi
List of Tables xiii
Series Foreword xv
Preface xvii
About the Authors xxv

Part 1. The Power of Writing: Evidence From Letters 1

1. *Writing as a Fundamental Mode of Communication* 3

2. *The Mechanisms of Writing* 17

3. *Expressing Emotions Through Writing* 31

4. *Knowledge Development Through Writing* 61

5. *Writing and Community Building* 81

Part 2. The Power of Writing in Online Communication 107

6. *From Letters to Online Writing* 109

7. *Expressing Emotions and Developing Trust Online* 121

8. *Creating Knowledge in Online Interactions* 141

9. *The Role of Writing in Developing a Sense of* We-*ness in Online Communities* 161

10. *Beyond the Media: The Power of Writing* 177

Notes 195

References 201

Index 217

List of Figures

Figure 3.1 Map of Hudson Bay and Straits "Made by Samuel Thornton at the Signe of the Platt in the Minories London Anno 1709." By Samuel Thornton. HBCA G.2/1 (N4605) 37

Figure 3.2 *York Factory* (1770s?), by Samuel Hearne. Engraving, colored, March 1, 1797. Engraver: Wise. HBCA P-228 (N5411) 38

Figure 3.3 *York Factory—Arrival of Hudson's Bay Company's Ship* [ca.1880]. Engraving (B&W & hand-colored) by Schell & Hogan. Taken from *Picturesque Canada* (Vol. 1, p. 317), 1882, Toronto. HBCA P-676 (N14407) 39

Figure 3.4 Letter from Moose Factory to London, dated August 17, 1739. HBCA A.11/43 fo. 19 (N16898) 40

Figure 3.5 Albany Fort Standard of Trade, 1706. HBCA B.3/d/15 fos. 9d-10 (T16889) 41

Figure 3.6 Portrait of Virginia Woolf, by Charles Beresford (1902) 51

Figure 4.1 Albert Einstein during a lecture in Vienna in 1921. Photography by Ferdinand Schmutzer (1870–1928) 73

Figure 4.2 Élie Cartan 74

Figure 5.1 Infographics of Mersenne's network. © 2010 by Guilhem Tamisier. The analysis was based on the data provided by Hans Bots (2005) in "Martin Mersenne, 'secretaire general' de la Republique des Lettres (1620–1648)," in Berkvens-Stevelinck, Bots, and Haseler (2005) 90

Figure 5.2 *Portrait of René Descartes* (1596–1650), oil on canvas, reproduction by André Hatala [e.a.] (1997) De eeuw van Rembrandt, Bruxelles: Crédit communal de Belgique, ISBN 2-908388-32-4 92

Figure 5.3 *Portrait of Émilie du Châtelet* (1706–1749), by Maurice Quentin de La Tour 1704–1788 96

Figure 5.4 *Portrait of François-Marie Arouet de Voltaire* (1694–1778), oil on canvas, by Catherine Lusurier (around 1753–1781) inspired by Nicolas de Largillière (1656–1746); 1778 (copy), 1718 (original) 97

Figure 5.5 *Portrait of Pierre Louis Maupertuis*, 18th-century drawing 98

List of Tables

Table 2.1 The mechanisms of writing 23

Table 3.1 Writing's mechanisms and their effects on emotional expression 34

Table 3.2 Expressing emotions in organizational letters 42

Table 3.3 Expressing emotions in personal letters 52

Table 4.1 Writing mechanisms and their role in knowledge sharing and development 63

Table 4.2 Development of an organizational memory through writing 67

Table 4.3 Development of a scientific theory through writing 75

Table 5.1 Writing mechanisms and community building 87

Series Foreword

In this very first organizational volume to carefully explore the role of the written word in organizations, Fayard and Metiu bring together an impressive set of historical and contemporary case studies to illustrate how the functions of organizations are intertwined with the mechanisms of writing. Fayard and Metiu show us how the organizational processes of knowledge sharing, wide-spread collaboration, and the communication of emotion—which are often thought of as thc domain of face-to-face interaction—are dependent on and influenced by written communication. These effects are evident in reading not only the letters of 18th-century commanders of the Hudson's Bay Company, but also in online exchanges of participants on OpenIDEO, the open innovation platform launched by the modern design company IDEO. Despite visions of a 21st century driven by video communication, Fayard and Metiu make clear how our current and future organizing is, and will continue to be, embedded in the power of writing.

Kimberly D. Elsbach,
Arthur P. Brief, and Michael Frese,
Series Editors

Preface

This book started as a conversation between the two of us, face to face and on the phone, but mostly through e-mails, while working at the same school. One was based in Asia, the other in Europe. Some of our conversations were about research topics we were both interested in—in particular, studies and debates in organizational studies, organizational communication, and information systems about the impact of new technologies on emotions, knowledge, and communities. We were both surprised by studies claiming that people could not collaborate, build relationships, or develop knowledge by means of written communication, while our research (along with others' research) and our experience showed the contrary. It is true that we, as have others, observed the "death by PowerPoint" phenomenon in organizations and in the classroom, with otherwise capable students unable to write and develop an argument. We also listened to executives and managers, whom we were teaching and consulting, voice similar complaints about their collaborators' lack of writing and analytical abilities and the negative consequences of this lack on performance.

At the same time, negative experiences using writing-based media to work with others (being overwhelmed with e-mails and the writers' expectations to receive an instantaneous reply; waiting for a reply to an urgent message; or getting short, brisk messages that we might perceive as unfriendly or even rude) cannot annihilate positive experiences (writing articles and designing complex research projects with others while rarely meeting or speaking, being members of enthusiastic online communities, or keeping in touch with friends and families while leaving abroad). Clearly, the potential of writing-based collaborations is there, as illustrated by numerous instances of successful collaborations in both informal and formal organizations, among which are the cases and the interviews presented in this book. But what is the potential lying in writing, what are its powers, and on what do they rest? And how do these powers play out in essential organizational processes? These were the questions that guided the inquiries that resulted in this book.

For us, the focus on writing as a fundamental mode of communication—the other being speech or the oral mode—was needed, even in our technology-infused world. Precisely because of their heavy reliance on technology, current practices are in need of a deeper understanding that would focus on deep, as opposed to surface, features. Yes, technology was increasingly ubiquitous in our personal and working life; yes, these new media afforded new practices (not that we had to necessarily buy into all of them); yes, their features kept changing, but at the end of the day, wasn't it all about writing? Don't all these media—an e-mail, a post on an online forum or on a blog—share something fundamental? Aren't they all instantiations of the same written mode of communication? Haven't humans been writing for millennia? Wasn't writing often described as one of humans' main inventions, opening new possibilities to the human mind and human societies? We thought of famous examples of writing: correspondences of famous novelists such as Virginia Woolf or famous scientists such as Albert Einstein as well as the Republic of Letters in the 17th and 18th centuries, which functioned as a scientific community distributed across Europe.

To find answers to our inquiries, we started reading some correspondences in informal as well as formal organizations—such as the Hudson's Bay Company, one of the ancestors of today's globally distributed organizations. We also turned to various sources of scholarship, not only the research in organizational studies, organizational communication, and information systems on contemporary communicative practices but also research by psychologists, historians, philosophers, and literacy scholars, who all have explored the role of writing for the human mind and literate societies. These latter studies confirmed our intuition and our empirical observations about how writing was crucial for expressing emotions, developing knowledge, and building communities—three areas that we call the powers of writing. Also, our reading of the literacy scholars led us to consider how some of the dimensions of writing mentioned by literacy theorists could support mechanisms central to the three powers of writing. We realized that while literacy theorists describe dimensions or characteristics of writing, it was more productive to analyze writing in terms of the mechanisms it afforded. Based on an inductive and iterative approach, and on historical analysis of written exchanges, we identified and defined four mechanisms of writing: objectifying, addressing, reflecting, and specifying. This is what we describe and discuss in Part 1. In Part 2, we use contemporary case studies and interviews and show how shifting our focus from the media to the mode of communication and focusing on the mechanisms allows us to go beyond current debates about the capabilities of various communication media (some claiming that relationships, knowledge, and communities are threatened by new media, while others are arguing that, on the contrary, new media are supporting distance collaboration and relationship, idea generation, and communities) and to better understand today's communicative practices.

This book demonstrates the power of writing in informal and formal organizations in the past and the present. It shows how writing, despite long-lasting criticisms that can be traced back to Plato and its frequent definition as solely a recording mode, is in fact a creative mode of communication supporting the development of reflection and critical thinking. Indeed, since its invention, writing has never ceased to allow scientists, philosophers, and novelists, and their communities, as well as members of organizations, to enact three essential powers of writing: expressing emotions, developing knowledge, and building communities.

Today's media do not always support writing mechanisms and, thus, their powers. Indeed, as people become more and more addicted to their smart phones and to having a constant Internet connection, some of the mechanisms of writing, such as reflecting or specifying, are often forgotten. Yet, these new media do not, by definition, prevent the enactment of writing's mechanisms. There are in fact many successful cases of online support communities, such as the WELL, and of distributed knowledge creation, such as open source software Wikipedia, and open innovation platforms. A close look at these cases using a mode perspective shows that the enactment of writing mechanisms is critical in these various settings.

These successful cases should not, however, make us underestimate the potential losses brought about by communication technologies. About two years into writing this book, we were pleased to discover that some of the issues we were addressing were the foci of interest of other researchers. Authors such as Maryanne Wolf, Nicholas Carr, and Sherry Turkle have started highlighting the potential losses for emotions, knowledge, and communities if we were to stop writing and reading and went into a spiral of constant interruptions, brief messages, and an endless flow of data that accompany new communication technologies, such as e-mails or Twitter. Carr and Turkle tend to focus on the negative aspect of the Internet, constant connectivity, and mobile devices. We found particularly enlightening Maryanne Wolf's work, which provides a deep analysis of reading and of the reading mind (reading and writing being deeply intertwined); she shows the incredible cognitive abilities reading and writing have provided human beings and reminds us to be aware of what we have and what we might lose if we stop nurturing and cultivating them.

While the work of these scholars is extremely informative, we believe that the dangers they identify, while quite real, are not fatal. We also believe that there is room for agency in the use of these technologies, both at the individual and the organizational level. While there is a collective agreement among practitioners and managers that we do not have time to read either newspapers, documents of more than two pages, or, we need not mention, novels, and while there is an agreement that, increasingly, people do not know how to write anymore, people take steps to counterbalance these tendencies. For example, companies send their employees to writing workshops, and the curriculum in New York public schools strongly emphasizes writing and reading as key processes for developing critical thinking.

In this book we attempt to unveil the powers of writing and to highlight the implications for organizations of the potential loss of these powers. We hope the book will be helpful to our peers, to scholars, and to students of organizational studies, organizational communication, and information systems who are investigating and theorizing about the communicative practices brought about by new communication technologies and their impact on organizational processes. We think it will be of use in particular to those examining knowledge sharing and development in organizations, investigating trust and emotions in distributed collaborations, or analyzing the effectiveness of technology-mediated interactions. We also hope it will be useful to the managers and other professionals who either sat in some of our classes or who we interviewed. We hope it will provide them with new ways to think of technology and communication as they struggle to balance the urge for effectiveness and reactivity in a world of distributed collaboration, ubiquitous computing, and high connectivity.

This book provides a review of the literature in diverse fields, such as literacy studies (sometimes also called classical studies), psychology, communication, organization studies, and management, on which we built our argument for the need to shift our analysis from the media and their features to writing as a mode of communication. This argument is grounded on our analysis of various case studies—some historical and some recent—of informal and formal organizations. Part 1 of this book, *The Power of Writing: Evidence From Letters,* presents the main debates about writing, develops our mode perspective in introducing the four mechanisms of writing (objectifying, addressing, reflecting, and specifying), and takes a historical perspective to illustrate the three powers of writing (expressing emotions, developing knowledge, and building communities) in several sets of correspondences: Virginia Wolf's letters with her family and friends, letters exchanged by the management of the Hudson's Bay Company and the post commanders in today's Canada, the correspondence between Albert Einstein and Élie Cartan, and correspondence among several members of the Republic of Letters.

In the second part, *The Power of Writing in Online Communication,* we show how our theory is a tool that is relevant to understanding communicative practices with online media through an analysis of current contexts, such as the open-source software development community and the OpenIDEO platform, and interviews with professionals in various fields, such as architecture, media, and publishing. Our analysis demonstrates how the enactment of the four *mechanisms* of writing is still possible and essential to the three *powers* of writing: expressing emotions, developing knowledge, and building communities.

This book has several distinctive features. First, it offers a powerful perspective on communicative practices by focusing on writing as a mode of communication instead of on the features of the media. This allows an examination of the causal mechanisms that support the powers of writing and provides a solid basis for the examination of the changes brought about by various media. Second, it provides a pertinent and rare combination of a historical analysis of correspondences from the 17th, 18th, 19th, and 20th centuries, with an analysis of contemporary practices in formal and informal organizations. The historical analysis allowed us to avoid being caught in futile examinations of the constant changes and multiple features of new media and thus to identify the stable mechanisms that characterize writing. Third, we take a multidisciplinary approach. Some scholars in others fields, such as classical studies, communication, psychology, or history, have provided rich insights on writing, highlighting some of writing's dimensions or characteristics and their importance for knowledge development and thinking, in particular. Yet, very few have attempted to provide a comprehensive multidisciplinary perspective on writing as it applies to fundamental processes such as emotional expression, knowledge development, and community building in formal and informal organizations. Fourth, the book provides an adequate balance between theoretical development and empirical work: It develops a new theoretical perspective on organizational communication, and it is grounded in analyses of numerous case studies.

PERSONAL ACKNOWLEDGMENTS

The sociologist Howard Becker (1984) taught us how art was a collaborative activity and not solely the work of "the artist." Becker's claim has been in our minds during the entire book project. Of course, the collaboration is clear in its coauthored nature, with us located on different continents, but this collaboration is more than a dialogue between the two of us: It includes a constant conversation with the series editor Kim Elsbach, who has been an enthusiastic supporter of our ideas since the beginning, before the notion of a book had crossed our minds. Kim's generous and critical feedback was instrumental in developing these ideas into a book form. Reviewers in different review processes, for the book as well as for articles and book chapters anterior to this book, are also key actors. Among them, JoAnne Yates, Beth Bechky, Wanda Orlikowski, Mark Zbaracki, and Myles Jackson have provided especially insightful and helpful suggestions. We also are extremely grateful to Ashwin Gopi, a student in management at NYU–Poly, who worked as a graduate assistant on this project and was incredibly effective in helping organize different parts of the book, such as the references, the images, and the copyrights as well as analyzing some of the data with us. We are grateful to Roan Pastor, who has helped us create a cover that reflects our thinking. We also would like to thank Guilhem Tamisier for his visualization of Mersenne's network and his inputs on the book cover. We are thankful to our interviewees for their generosity imparting their experiences with, views on, enthusiasms about, and struggles with the written word and the wild world of the new media. We thank Anne Duffy and the other Taylor and Francis employees who provided steady support for the book.

Anne-Laure Fayard. This book started long ago, as I learnt to write, to master grammar and spelling, but most of all, to articulate a thesis and develop an argument. In that apprenticeship, many teachers and professors were involved. My readings of philosophers, in particular Plato and Descartes, were central to my understanding of the writing process. I remember lively discussions with Jacques Deschamp as I first read these authors. My family has been very patient and passionate about this book. I am grateful to their curiosity, encouragement, and support. Melchior and Jyoti learned enthusiastically about Madame du Châtelet, Virginia Woolf, and King Hammurabi, among others, as well as about the importance of writing. Guilhem has never grown tired of our discussions about writing, letters, and new media. Many days when I worked from home, or during evenings and weekends, my cat, Noa, was a wonderful companion, who sadly left us before the book was finished. Many friends kindly listened to, or read about, our progress and frustrations. I dearly value their continuous support, even though, at times, it seemed to them a never-ending task. Foremost, I thank Anca for embarking with me on this exciting adventure: Skype or phone calls, face-to-face meetings during the summers, innumerable e-mails, multiple versions and revisions, and now, a book!

Anca Metiu. This book is my alms to the writers—alive and dead—who have taught me the unequaled power of the written word. To Proust and Dostoyevsky and Woolf and Kafka and Faulkner and Kundera and Morrison, who have created worlds as real as anything one can brush her cheek against. To my writer friends, who are at this very moment struggling to extract similar potent elixirs from words and their arrangement on a sheet of paper or on a blog. To Mr. Fantana, my first literature teacher, whose assertion that linking words—prepositions and conjunctions—are even more important than nouns and verbs was intriguing enough to stay with me for all these years and to find an echo in my work. To Anne-Laure, my writing partner, for her boundless enthusiasm, energy, and urgency. To Alma and Jacques, for making it all worthwhile.

ACKNOWLEDGMENTS FOR SOURCES

Quotes from *Congenial Spirits: The Selected Letters of Virginia Woolf,* 1989, edited by J. Trautmann Banks, published by Hogarth Press. Used by permission of The Random House Group Ltd and Houghton Mifflin Harcourt.

Quotes from "Letters from Hudson Bay 1703–1740." In *Hudson's Bay Record Society, Vol. 25,* 1965, edited by K.G. Davies and A.M. Johnson, and images from Hudson's Bay Company. Used by permission of The Hudson's Bay Company Archives (HBCA).

Quotes from *Élie Cartan–Albert Einstein, Letters on Absolute Parallelism, 1929–1932,* edited by Robert Debever © 1979 by The Estate of Albert Einstein and the Estate of Élie Cartan. Used by permission of Princeton University Press.

Quotes from *Lettres Innédites de Madame du Châtelet et supplément à la correspondence de Voltaire avec le roi de Prusse et avec différentes personnes célèbres, Imprimerie de Lefebvre, Paris 1818*. Source: Bibliothèque Nationale de France BnF/Gallica and Googlebook. In the public domain.
Book cover image © 2012 by Roan Pastor.
Infographics of Mersenne's network © 2010 by Guilhem Tamisier.
Some of the analyses presented in this book are grounded on previous studies published in articles and book chapters, by the two authors, together and separately. In each specific chapter, the original work is referenced, but we would like to acknowledge these sources:
Fayard, A. L., & DeSanctis, G. (2010). Enacting language games: The development of a sense of "we-ness" in online forums. *Information Systems Journal, 20*(4), 383–416.
Fayard, A.L., & Metiu, A. (2009). Expressing emotions and building relationships over distance: Fixedness and fictionalization in correspondence. In K. Elsbach & B. Bechky (Eds.), *Qualitative organizational research: Best papers from the Davis Conference* (pp. 49–180). Greenwich, CT: Information Age.
Metiu, A. (2010). Gift-giving, transnational communities, and skill building in developing countries: The case of free/open source software. In M.-L. Djelic & S. Quack (Eds.), *Transnational communities and the regulation of business* (pp. 199–225). Cambridge: Cambridge University Press.

About the Authors

Anne-Laure Fayard earned a B.A. in Philosophy from La Sorbonne–Paris I University in 1991, an M. Phil. in History and Philosophy of Science at La Sorbonne–Paris I University in 1993, an M.Phil. in Cognitive Science at the Center of Applied Epistemology (CREA) at the École Polytechnique (Paris, France), and a PhD in cognitive science from École des Hautes Études en Sciences Sociales (Paris) in 1998. Since that time, she has taught at INSEAD International Business School, both in Fontainebleau (France) and in Singapore, prior to joining the faculty at the Polytechnic Institute of New York University in 2006. She has held visiting positions at the Center of Sociology of Innovation at École des Mines in Paris and at Design London and the Innovation and Entrepreneurship Group at the Imperial College Business School in London.

Her research interests include organizational communication (including technology-mediated communication) and sociomaterial practices. She has examined the role of space—physical and virtual—in triggering informal interactions and supporting collaboration as well as the enactment of language games and discursive practices in online communities. Some of her recent work investigates the sociomaterial practices developed by open innovation intermediaries, and the practices enacted by service designers in designing services and constructing an occupational mandate. Her research has been published in articles and book chapters in a variety of prestigious outlets, including *Organization Studies,* the *Harvard Business Review,* and the *Journal for the Association of Information Systems, Information Systems Journal.*

She regularly reviews for numerous journals in the fields of management and information systems, including the *Organization Science, Organization Studies, Journal of Management Inquiry, Information Systems Research, Management of Information Systems Quarterly,* and *Human Relations.* She is a member of the Academy of Management, the European Group of Organization Studies and the Association of Computing Machinery.

Anca Metiu earned a B.A. in Law and Economics from the University of Sibiu, an MBA from the University of Illinois at Urbana–Champaign, and a PhD in Management from the University of Pennsylvania. Prior to joining the faculty at ESSEC Business School in 2007, she taught at INSEAD.

Her research focuses on collaboration dynamics in distributed work. She has examined the status dynamics among distributed work groups and the work organization of developers in the free and open-source software development community. Some of her current projects examine the creation of group engagement in teams, the perceived proximity between people who work across geographic distance, and the processes of professional identity formation for women in the free/open-source software communities.

She has published articles and book chapters in a variety of prestigious outlets, including *Administrative Science Quarterly, Organization Science, Organization Studies,* and the *Oxford Review of Economic Policy.* She is currently serving on the editorial board of influential journals in the management field, including *Organization Science* and *Organization Studies.* She is a member of the Academy of Management, American Sociological Association, European Group of Organization Studies, and The European Academy of Management.

Anca Metiu teaches in the executive, master's, MBA, and PhD programs and coordinates the management track in the PhD program at ESSEC Business School.

Part 1

The Power of Writing

Evidence From Letters

CHAPTER

1

Writing as a Fundamental Mode of Communication

What would a world without writing look like? It is hard, even impossible, to imagine. It would be a world without novels and literature as we know it, without administrations and complex organizations, without science, philosophy, and history. It is easy to see then why writing is widely considered one of the most important inventions of mankind: It provides a foundation for our societies as well as for our ways of thinking. In fact, oral cultures (those that rely purely on oral tradition, such as the Homeric Greeks, the Lakota Sioux in North America, or the Mande in West Africa) are nowadays difficult to imagine because purely oral cultures[1] no longer exist, and when we think of words, we visualize them as written symbols to be decoded.

This book is about writing and its powers, and about how this mode of communication has allowed us—it still does!—to express emotions, develop knowledge, and build communities. While many scholars focus on the changes introduced by the new media and their constantly evolving features and engage in debates about whether these new media allow the expression of emotions, the development of knowledge, and the building of communities, we suggest stepping back and focusing on writing as a mode of communication (the other mode of communication being oral). Our analyses of historical correspondences and of more recent communicative practices in organizations and distributed communities follow in the footsteps of literacy scholars and psychologists who have shown that writing is intrinsically creative. We identify and examine four mechanisms—objectifying, addressing, reflecting, and specifying—that are at the core of the writing process and give writing its powers: the power of expressing emotions, of developing knowledge, and of building communities.

Before providing a detailed outline of the book, and to provide context for our argument, we examine the role of writing in organizations and society as well as highlight opposing views on that role. We show that while writing has been central to the development of organizations, scientific knowledge, and communities, people tend to consider it as limited compared to face-to-face interaction, which is often seen as the ideal form of communication. We provide a brief review of different opinions about writing and use contemporary examples of reliance on writing to introduce the idea that writing is a powerful mode of communication, not to be mixed up with the various media in which it is performed. We then outline the main arguments and structure of the book.

THE ROLE OF WRITING IN ORGANIZATIONS AND SOCIETY

Writing and organizing have always been intrinsically linked. Indeed, writing was invented approximately 5,000 years ago in the earliest cities and states of Mesopotamia, where leaders' governance was challenged by population growth: Their societies had escalated from a few hundred villagers to tens of thousands of city dwellers. Such communities could not be run by word of mouth, and most of these new leaders found that, besides maintaining consistent political ideologies and strong militaries, they needed to write things down. Thus, the first examples of writing (such as on tablets in Mesopotamia) did not offer poetry (early literature was oral and learned by heart) or science but rather records of state goods. People wrote what they could not remember: records, for instance—such as which workers had been given their allotment of beer and other accounting matters.[2] King Hammurabi, who ruled Babylon in ancient Mesopotamia, is famous for being the first king to write laws that could be referenced and known by all. Thus, early writing was not about emotions or science, but about money, laws, trade, or employment. Its development changed the nature of state control and state power from one based on military might to one based on bureaucracy and economy. As recently noted by Neil MacGregor, the director of the British Museum, in a BBC program, *The First Cities and States: A History of the World*, by 3000 BC, managers of Mesopotamian city-states were able to use written records for administration purposes such as keeping large temples running, managing goods, and so forth.[3]

Ever since its invention, written communication has played a crucial role in organizations' growth and expansion across the globe (King & Frost, 2002; Yates, 1989). Because it leaves a trace, writing enabled record keeping, knowledge sharing, and the coordination of complex activities spanning a range of specialties and regions (King & Frost, 2002). In particular, writing fixed meaning, which enabled members of an organization or a community to share information and ideas they could refer to without having to endlessly frame the messages (King & Frost, 2002).

As a result, it is difficult to picture what an organization, let alone society, would look like without writing. Nowadays, we live in literate cultures (Ong, 1982/2002), with societies, states, and organizations all relying on writing for records such as birth certificates, diplomas, bills, invoices, and taxes. Our literature has become writing based, so much so that oral literature has become nearly an oxymoron. Even talks or speeches are first written to be later read in public. Furthermore, the complexity of our knowledge (the sciences, philosophy, social sciences) makes us dependent on writing, not only for preserving existing knowledge but also for creating it (Searle, 2010). While the medium has changed—from clay tablets in ancient Mesopotamia to papyrus in ancient Egypt, parchments during the Middle Ages, or letters in the Renaissance—we still write a lot: from short notes and e-mail (or even text messages on the phone) to long, formal, and complex e-mail that resembles the letters of a bygone era. In fact, the importance of writing for modern organizations continues unabated, with e-mail becoming indispensable (Byron, 2008; Gardner, 2008).

In the following two sections, we will first discuss the tension between writing and speaking and, more specifically, the long-standing contradictory imagination about writing and the presupposed superiority of oral communication. We will then explain how switching the focus of analysis (from the medium to the mode of communication) and the analytical approach (from exclusively contemporary developments to a more encompassing historical perspective) lets us identify the specific mechanisms underlying the powers of writing.

Contradictory Imaginations About Writing: The Tension Between Writing and Speaking

Despite claims that establish writing as one of the greatest inventions of mankind, and despite its central role in the development of our societies and our daily reliance on it, a tacit assumption persists: that face-to-face interaction is the ideal, richest form of communication and that nothing can replace it when it comes to creating knowledge, expressing emotions, or being part of a community. Although we spend a lot of time typing on the computer (reading newspapers, books, papers, e-mail, or searching the Web), and despite the fact that many of our oral interactions are not face to face but rather voice to voice (on the phone or voice calls on the Internet with software such as Skype), we tend to perceive face-to-face communication as "truer" and deeper, more authentic and more genuine. This assumption about face-to-face communication endures in spite of the general discourse about virtual organizations, virtual communities, virtual worlds, and distributed work, which claims that we can work across geographies and spaces as well as build relationships with people we have never seen and will probably never meet. In fact, this assumption led to huge investments during the 1980s in video conferencing systems that would recreate face-to-face interaction and overcome distance. Yet these investments led to mitigated successes, and many companies these days use teleconference calls and a massive amount of e-mail and other shared, text-based documents that are annotated and collaboratively modified. Even phones, first seen as representative of oral communication, are now increasingly used to send text messages instead. In 2009, more than 1.5 trillion text messages were sent or received in the United States.[4] In the same year, in France, about 62 billion text messages were exchanged.[5]

Thus, we are faced with a tension between recognizing that writing is a key invention and that our cultures are literate and the belief that face-to-face interaction is ideal and that we should therefore try to reproduce it. This tension reflects a conflicting imagination about writing. We fantasize that we can move away from it, even as we do it more and more, as if we were condemned to write, while we dream of not writing and not reading, just talking, as if writing were second best to talking (or good for record keeping, at most), with face-to-face dialogue being the only way to convey or develop "true" knowledge, "true" emotion, or "true" bonds. The phrase "A picture is worth a thousand words" exemplifies that belief. Written words are seen as "dead signs" that freeze—even kill—creativity.

The contradictory imagination about writing and the presupposed richness of oral communication is not new. Since its invention, writing has been held inferior to the spoken word. Plato, who is considered the first philosopher to write, was a strong critic of writing, and he argued that words on a page are dead things that cannot speak, answer questions, or come to their own defense (e.g., *The Phaedrus* 275e [Plato, 1940a]). Plato's criticism of writing bears a striking resemblance to current objections raised by media-richness scholars (Daft & Lengel, 1984, 1986). For example, today's scholars criticize writing for its inability to provide immediate feedback, just as Plato criticized writing for simply being "dead signs." He also criticized writing because he believed it prevented people from remembering the archetypal Ideas that constitute true knowledge.[6] That is, because writing produces dead signs, it can only help those who already *know* to remember, but it cannot teach them anything. This criticism is similar to current objections about the side effects that Googling has on our memory capacities and is reminiscent of older objections that arose when the printing press and telegraph were invented. In fact, over the past 150 years, the advent of each new communication medium has

HISTORICAL CRITICISMS OF WRITING

The numerous criticisms of writing formulated over time can be reduced to three main traditions that consider written words as dead signs on a page. (For a detailed review, see Bazerman, 1988, pp. 19–24.) The first criticism dates back to Plato and argues that meaning lies in primary referents, or a world of essences, outside the symbols used to clothe them in texts and that one can find meaning only in the philosophical dialogue. As noted by Bazerman, although we might disregard the idea of a world of essences, this criticism can resonate with our frustrating attempts to capture in written text our thoughts and feelings, only to be thwarted in our attempts to "reach" the meaning we're trying to convey. This struggle highlights the creative and processual nature of writing through which writers try to match their ideas or emotions with the symbolic system of language.

The second criticism, developed originally by the Sophists, and more recently by structuralists and deconstructionists, purports that the meaning of the text is enclosed entirely within the text, which is an arbitrary cultural system. According to these critics, the text is a system that defines the meaning; without the socially defined constraints of our linguistic system, it could mean anything. This extreme relativism, where language can mean anything once the rules of the game are defined, is tested by examples of translations and collaboration across different groups and societies. Such examples indicate that some common elements of life have a similar meaning despite their expression in different linguistic systems.

The third criticism defines language as a socially constructed system and considers written language, which lacks social context, as an epiphenomenon, a pale reproduction of the living (spoken) language. This corresponds to the belief that communication and community arose from interactive, face-to-face dialogue. This criticism highlights the importance of the social context and specific situations to understand communication. However, as Bazerman notes, writing is far from a decontextualized process: "Writing and reading may take place in privacy and composure, and they may carry out distant social actions, but they are still highly contextualized social actions, speaking very directly to social context and social goals" (Bazerman, 1988, p. 23). This once again refers to the importance of writing as an interaction (between the writer and his or her audience) and as a situated process. Even if written letters or books appear to transcend geographies and times, they are still contextualized—intentionally written for someone and always interpreted.

elicited both enthusiasm and criticism (Gardner, 2008). (See the sidebar Historical Criticisms of Writing for a short overview.)

These similarities between past and current criticisms of written communication suggest that contemporary debates about the capabilities of various media to convey ideas and emotions are but the latest iteration of the deeper tension between the oral and the written. They also hint that, if we want to understand the powers of writing and their underlying mechanisms, we need to bracket the medium of communication to avoid being caught in a battle about a specific technology and its features. We do this in our book by analyzing correspondences and by focusing on writing as a mode of communication rather than on its embodiment in various technologies.

This change in perspective may seem odd at a time when the development of communication technologies is accompanied by a belief in technology's ability to surmount the shortcomings of distance and writing. This belief has led to a world inundated by videos and images, with people claiming they no longer have time to read and that no document should be more than a page long. Newspapers such as the *New York Times* or the *Guardian* increasingly use photo galleries, videos, and infographics to present their content in an attempt to attract "readers" who "don't have the time" to read. Managers tell their employees to write no more than one page for project proposals because people won't read more than that and to create PowerPoint presentations that rely on bullet-pointed statements rather than reports or documents filled with "bothersome" sentences. The increase in various forms of visual representations—videos, charts, and infographics—often associated with synchronous communication, has sometimes been described as "a second orality" (Bolter, 2001). This belief in a return to orality posits writing as a mere recording technology, a mode that is potentially outdated and can be replaced by images, videos, and sound.

The power of writing is intact, though, in our technology-infused world. A few years ago, as we were writing about the correspondences we describe in later chapters, we read an article in the *New York Times* (Seligman, 2009, p. A19) that perfectly illustrated the contradictory imagination that writing is an impoverished yet deeply powerful way of communicating. The article was written by the wife of a U.S. soldier, describing how her husband went away to war and left her alone with first one child, then two, and how after trying phone and Skype—because these technologies, they believed, would allow them to feel closer and share more feelings—they went back to the "good old letters." She wrote:

> We talked—sometimes twice a day—ignoring the popping and snapping on the line and the long delays between our voices on the Webcam. And I fooled myself into believing a two-dimensional image could transmit and sustain a three-dimensional marriage. After all, I could see his eyes, hear his laughter. But he knew nothing of what I thought about our marriage, nothing of my postpartum depression and nothing of my anger at feeling lonely in a life that he chose.

> How could I look at him on the Webcam and tell his sad eyes that I felt abandoned? How would I live with myself if, God forbid, the last words he heard from me were painful truths? The pressure to keep our conversations light controlled me, and it brought our marriage to a halt. When he returned from Afghanistan, I almost left him.
>
> The second time, when he left for Iraq, he said: "Let's not make the same mistakes.... No secrets this time." I nodded, even though I knew full well that, faced with the Webcam, I would again hide my fears and anger. (Seligman, 2009, p. A19)

As their relationship was dragging along, she decided to write him letters based on her diary, and this allowed them to exchange complex emotions with details and subtlety:

> And then we found salvation in letters. I had always kept a diary, but growing frustrated with my inability to really connect with David through the Webcam and on the phone, I started sending him long letters from my journal. Before long, I was picking out stationery to match my moods and searching for the perfect pen to carry my thoughts. David responded with enthusiasm. (p. A19)

This story shows how writing, despite its asynchronous nature, or maybe because of its asynchronous nature—so often regarded as a flaw—can support the expression of deep and strong emotions and support the maintenance (here, the rebuilding) of a relationship.

> Writing allowed us to regain control of our marriage. On paper, our memories came to life. Through letters we could share our concerns without worrying that we'd be misinterpreted. The children were able to keep their dad's letter in their pocket or drawings he sent them as a symbol of his love for them and they sent him drawings that he cherished. (p. A19)

This example, however, opposes two assumptions often taken for granted: first, that writing cannot support distant relationships, and second, that emotions cannot be expressed through writing and, by extension, that audio or video (e.g., Skype) is better because face-to-face exchanges offer synchronous communication, as well as verbal and nonverbal cues. Indeed, in this case, several dimensions of writing played an important role. It was because writing is asynchronous that this couple was able to analyze their feelings, express them in detail, and reflect on their own emotions as well as on each other's before responding. The delay offered by writing also allowed them to be more "truthful." When using Skype, they would try to look happy and positive, even if they felt the opposite. In the relationship created by writing, however, they were able to express negative and contradictory emotions. Moreover, as they were separated, they were able to imagine each other and become more deeply aware of the complexity of their emotions and of their relationship than they were able to in the video interaction over Skype or during a phone conversation. Finally, it mattered that the physical letters could be kept and carried around; they became literal symbols of their relationship.

This example recalls famous examples of letters exchanged between lovers, friends, and families across the centuries, such as the letters of poets Elizabeth Barrett and Robert Browning, who fell in love before meeting each other, or those of Franz Kafka and Milena Jesenská. The story is also similar to those told during the Victorian period to support the postal reform of one-penny letters (also known as the Uniform Penny Post). As Catherine J. Golden described in *Posting It: The Victorian Revolution in Letter Writing,* supporters of postal reform suggested that letters would allow friends and family who lived in different parts of the country and the world to maintain strong relationships at a time when British society had become increasingly mobile.

The example of the soldier and his wife is very strong, but can writing play a similar role in work relationships? We argue that it can, not only for expressing emotions that are increasingly recognized as crucial to organizations but also for the sharing and creation of new knowledge. Let us take the dramatic example of the failure of the Columbia space shuttle. In its report, the Columbia Accident Investigation Board argued that the use of PowerPoint and its hierarchical bullet-point format had a negative impact on the analysis of the situation and the decision making of managers (cited in Tufte, 2006). Lower level NASA employees, however, wrote about the possible dangers to the Columbia space shuttle in several hundred e-mail messages, 90% of which used sentences sequentially ordered in paragraphs. Tufte and others show how detailed, well-argued e-mail supported the analysis of the information and the sharing of crucial knowledge while PowerPoint, with its more visual format, failed to convey the available analyses and how that failure led to a major catastrophe (Tufte, 2006, p. 12).

Let us now consider a less dramatic context: smart phones in the workplace. In 2009, the *New York Times* (Williams, 2009) reported the debate in organizations about the use of smart phones in meetings: Are people rude and not committed to the meeting if they check their phones, or are they just trying to be efficient? Some argued that managers did not have a choice and that their clients assume they should be responsive 24 hours a day. Others admitted that they used smart phones to exchange comments with others about the meeting, creating a parallel, simultaneous conversation. The debate illustrates the impact of mobile devices on our communicative practices as well as the conditions in which these messages are written: fast-paced, high-pressure situations, where we are often trying to hide that we are reading or responding to a message. New devices and the conditions of their use have major consequences for writing. Messages exchanged in such contexts tend to possess a curtness and lack of reflection that frustrates recipients, many of whom feel that these e-mail messages are sometimes insulting in their tone and style. Moreover, such messages often fail to provide useful information for the problem at hand. This illustrates how writing, because of context and interface, can change its nature, becoming in some ways "oral" and losing the asynchronous and reflective dimensions traditionally associated with it. A good example of this development is the distinction people make between e-mail messages that are "written" (similar to letters or memos that contain information or arguments) and e-mail messages (or phone text messages) that are "chat"—for example, "almost there," "on my way," "running 10 minutes late." The blurred distinction between the oral and the written brought on by the influence of various communication technologies and devices is another facet of the "return to orality" (Bolter, 2001).

Current technological developments raise the fundamental issue of the power of writing. How does its historical role hold up in today's media-saturated world? Is it possible to find some solid theoretical ground for the examination of the current, rapid changes in the way people communicate?

FROM THE MEDIA TO WRITING AS A MODE OF COMMUNICATION

We start from the contemporary belief in the power of oral communication and from the historical fact that, for millennia, writing (and, particularly, letter writing) has allowed humans to develop and manage organizations, create knowledge and scientific theories, develop and maintain close relationships, and create communities. We propose a reframing of organizational communication that focuses on the *mode* of communication—written or oral, rather than attending to the *media* of communication (e.g., e-mail, text messaging, instant messaging, blogs). These are the two main modes of communication, and they are enacted in a variety of technologies, media, and devices. We take a mode perspective and focus specifically on the writing mode, because its role has not been properly acknowledged in organizational studies. Indeed, as the proliferation of new communication technologies, such as e-mail, blogs, social networks, and mobile devices, visibly influence the nature of organizational communication, they also trigger various debates among communication technology scholars about the capabilities or powers of various forms of communication (e.g., Byron, 2008; Daft & Lengel, 1986; Sproull & Kiesler, 1986). Overall, writing is viewed as an impoverished medium—as compared to the informal conversations at the water cooler or the copy machine—for expressing deep and subtle emotions or for developing relationships and sharing knowledge. In fact, many of the studies of organizational communication tend to focus on the features of the media and their influence on communication, while ignoring the fact that, despite the novelty of the medium, online interactions rely on writing, a fundamental and ancient mode. In our view, analyzing organizational discourse at the level of the mode provides a better understanding of its role in key organizing processes such as expressing emotions and developing knowledge.

Drawing on communication and literacy theories (Bolter, 2001; Goody, 1987; Havelock, 1963; Ong, 1982/2002) as well as research in cognitive psychology (Wolf, 2008), we suggest that it is productive to think about current communicative practices by accounting for their essential nature as written communication and to analyze them not only in terms of technology (e.g., quill pen, pencil, typewriter), media (e.g., telephone, e-mail, slide presentations, instant messaging), and devices (e.g., laptop, desktop, smart phone) but also in terms of modes of communication, whether oral or written. Even the latest communication media rely on these two fundamental modes of communication. In fact, new technologies tend to include a mix of audio, text, and writing-based forms of communication, sometimes including video. Skype, for example, offers voice calls on the Internet but also video, as well as instant messaging. Increasingly, people use their mobile phones not only to make calls but also to send text messages and even check their e-mail. While each mode of communication can be performed with various media—each of which possesses its own affordances and uses—distinguishing between modes of communication and media is productive because it offers a framework for analyzing today's communication practices, with their mix of oral and written components and with their ever-evolving features and capabilities.

Writing has been described as a technology—one of the major technological inventions that humans have developed. However, this can be confusing, as the term *technology* has come to be associated with the various media of communication or with the devices we use to communicate. In today's world, many kinds of media are available. We can communicate with people all over the world via e-mail, online forums, or instant messaging. We can share our views on blogs and electronic mailing lists. We can also follow people on Twitter, connect on LinkedIn, and make friends on Facebook. All these media have features of their own, and using them to try to understand how people communicate is nearly impossible, as technology keeps changing and features are constantly added. Thus, we conceptualize writing not only as a technology but primarily as a fundamental mode of communication. To avoid the continuous changes, and at the risk of looking old fashioned, we started by analyzing writing in the context of a simple, old-fashioned genre: letter writing. At first, letters may seem a peculiar type of written material, bearing few similarities to e-mail or postings on online forums, intranets, or blogs. But the differences in the media are overridden by the fact that these forms of communication are written and thus enact the mechanisms we identify in our analysis of letters.

This conceptual shift—from media to mode of communication—allows us to examine the power of writing and the communicative practices supporting better relationship building and knowledge sharing in organizations. In addition, conceiving organizational communication at the mode level exposes the conflicting views on the powers of writing and uncovers a set of mechanisms for the analysis of all written material, regardless of the medium in which the materials are enacted. It thus opens up rich avenues for empirical research on the consequences of new devices and communication media as well as the practices associated with them.

Acknowledging the power of writing is opportune in a time of increased mobility and connectivity, particularly for organizations where communication is central (Yates & Orlikowski, 1992). First, as organizations become more distributed, people work daily with colleagues, clients, and vendors in different time zones and geographies. Often they haven't met and might never meet, but it is critical that they collaborate efficiently. Being able to share emotions thus becomes essential, because doing so allows people to build trust and communicate effectively. Once again, although people might call each other, conduct video conferences, or meet face to face, many of these workplace interactions take place via writing. According to some counting and extrapolations by the Radicati Group, in 2010 about 9 trillion e-mail messages (not counting spam and viruses) were sent (Radicati, 2010).[7]

Second, as the speed of innovation quickens and problems grow more complex in today's knowledge-intensive and global economy, knowledge sharing becomes more crucial to organizational success (Powell & Snellman, 2004). As companies struggle with complex innovation problems, they reach out to a vast and diverse crowd of virtually distributed participants to not only share their knowledge but also to develop new, creative ideas; for example, crowdsourcing marketplaces such as InnoCentive and TopCoder or open innovation communities such as OpenIDEO are flourishing. Also, citizen science efforts, such as Galaxy Zoo. org have been publicized as the future of open innovation. In these open innovation platforms, most of the interactions are written. Photos, graphics, and videos are sometimes posted, but the primary mode for creating and sharing knowledge is through the written word.

Third, with nearly 2 billion people having access to the Internet, notions of community, public and private, are changing. Lam (1998), of the Pacific News Service, described how a majority of the 2.5 million Vietnamese displaced by the Vietnam War, who now live throughout the world, constructed "a Virtual Vietnam," where they share history, culture, and news, as well as establish social links. The open source software development movement, with its thousands of developers in various countries working together to produce free and accessible software, is another example of how people, through writing, are transcending distance to create and share knowledge and to build a community.

Because organizations increasingly rely on distributed work and, more recently, on open innovation platforms, as well as because of the emergence of virtual communities, such as open source software developers and citizen science projects, it is important to understand the role writing (the main mode of communication in these communities) plays in creating knowledge and building relationships. In this book, we argue that writing enables people in organizations to express emotions and build relationships, share and develop knowledge, and develop and engage in communities. We refer to these three processes—expressing emotions, developing knowledge, and building communities—as the three powers of writing.

BOOK STRUCTURE

We have been thinking about writing because our research and teaching compels us to do so, but also because the way we communicate with faraway colleagues, family members, and friends prompted us to reflect on all the things we can achieve while writing to distant others. This book is the product of many exchanges, short and long. We discussed our ideas through e-mail that at other times would have been letters, and we exchanged annotated manuscripts. It is likely that we write more e-mail than we would have written letters; yet, some of our e-mail messages are as intricate and complex as letters exchanged long ago. Our interest in developing an understanding of the role of writing as a fundamental communication mode led us to examine two sets of writing practices: letter writing, which we address in Part 1, as well as more recent writing forms, which we address in Part 2. As shown by scholars who studied the antecedents of modern management (Barley & Kunda, 1992; Guillen, 1997; Yates, 1989), a historical perspective allows us to better grasp the underlying mechanisms of writing, regardless of the media in which the modality has been enacted over the centuries. We therefore started from the analysis of the mechanisms of writing that emerged from the qualitative discourse analysis of several correspondences—the Hudson Bay Company, Virginia Woolf, Albert Einstein and Élie Cartan, Marin Mersenne, and finally Madame du Châtelet—and these mechanisms informed our analysis of contemporary communicative practices.

Of course there are differences between writing e-mail and a letter. Our written practices constantly evolve due to the emergence of new media and new interfaces. Yet some e-mail messages are not so different from classical letters in length, complexity, level of nuance, and styles. Furthermore, while we typically associate letters with complexity and nuance (perhaps because those are the letters we keep, examine, and study), we should not forget the many genres of letters: short notes, long letters, and informal as well as formal letters. These forms can, in some ways, be found in the variety of e-mail we send.

We also recognize the importance of understanding the various sociomaterial practices triggered and enacted by different technologies and media in a range of sociohistorical contexts. Think of the materiality of writing experiences and their social implications. Writing on a clay tablet with a stylus is vastly different from writing on a papyrus scroll with reed brushes, on parchment with a quill, or on paper with a pen. Those material writing practices are also radically different from typing on a typewriter, typing on a computer, or typing on a smart phone.

Not only is the experience materially different, but it has important social implications. For example, writing was for many years reserved to a small number—Egyptian scribes or medieval monks, say. Similarly, the invention of the typewriter greatly reduced mistakes in organizations because typed texts were much more readable than handwriting, and therefore good penmanship stopped being a prerequisite for some administrative jobs.

We focus on several historical periods that involved writing with a quill, writing with a pen, and typing on a typewriter or computer keyboard. This approach allows us to explore writing elements enacted through different correspondences at different historical periods. This historical perspective illuminates unchangeable dimensions of the writing mode, despite developments in the sociomaterial practices through which writing has been enacted: from letters in the 17th century up to now in online forums, blogs, and e-mail.

In sum, we propose a new way of analyzing organizational communication, from an emphasis on the media to an emphasis on the mode. We focus specifically on the written mode, which has not been properly acknowledged in organizational studies. We argue that focusing on the modes of communication allows us to better understand the role of writing in key organizing processes, such as emotional communication, knowledge development, and facilitating participation to communities. The approach also prompts better analysis of the changes that result from the introduction of new communication technologies.

Part 1 presents a historical analysis of letters in a variety of settings: distributed organizations, scientific communities, and writers and their family and friends. Chapter 2 discusses the theoretical background of the book and the four mechanisms of writing—objectifying, addressing, specifying, and reflecting—that we identified based on the discourse analyses of correspondences. Subsequent chapters present evidence that supports our claims. In Chapter 3 we examine two sets of correspondence (Virginia Woolf's letters with her family and friends, and the letters exchanged by the management of the Hudson Bay Company and the post commanders in today's Canada) and explore how emotions can be expressed in writing. In Chapter 4 we analyze the role of writing in the development of a new theory and of organizational memory by drawing on the correspondence between Albert Einstein and Élie Cartan, and on the Hudson Bay Company letters. In Chapter 5 we examine the role of writing in the building of communities, in particular, the development of the Republic of Letters.

After examining the power of writing in letters exchanged in formal and informal organizations in Part 1, we demonstrate in Part 2 the strength of our approach for understanding current communicative practices. In the second part we examine, through the lens provided by the four mechanisms of writing, writings' powers as used in current collaborations—again, both in formal and informal organizational settings. Based on interviews in organizations and case studies of online communities, we identify the changes introduced by new media and the resulting consequences for expressing emotions, knowledge development, and community building. Chapter 6 revisits the four mechanisms in the light of the new media and the changes they bring in communication practices and organizations. Chapter 7 analyzes the way emotions are expressed in contemporary mediated settings, in particular, with the case of "flaming" messages (hostile and insulting between participants) in open source projects. Chapter 8 shows how our ideas help explain the use and limitations of new media and technologies in some of the most recent forms of collaborations for knowledge development: collaborative design of concepts for social innovation challenges on OpenIDEO. Chapter 9 shows how writing-based modern technologies open up other avenues for the development of online communities and examines the development of a sense of "*we*-ness" or shared identity in a public online forum. Finally, in Chapter 10, we discuss the theoretical and practical implications of our perspective for researchers and managers grappling with key organizational processes, such as communication, creativity, and innovation.

CHAPTER

2

The Mechanisms of Writing

As we suggested in the previous chapter, shifting our focus from the characteristics of the media (e.g., pen and paper, print, e-mail, blogs) to the mode of communication—whether written or oral—allows us to analyze writing as a creative mode. This shift is inspired by the work of scholars such as Eric Havelock, the famous British classicist who proposed the thought-provoking and often controversial argument that the development of the Greek alphabet, and the related transition from oral to literate forms of culture, led to a dramatic change in the nature of ideas available to humans. After reviewing the work of such scholars as Havelock, Ong, and Goody, which is often coined "the literacy perspective," we present the four mechanisms of writing—objectifying, addressing, reflecting, and specifying—that emerge from the analysis of historical correspondences and that are, we argue, central to the creative nature of writing. More specifically, we claim that, despite debates in the current literature on organizational communication and media about the capabilities of writing, these four mechanisms support emotional expression, knowledge development, and the building of communities.

WRITING AS A CREATIVE MODE

Despite the usual association of writing with recording and reproduction, classicists, such as Havelock and Ong, psychologists, such as Olson and Wolf, and philosophers, such as Searle, argue that writing is, in fact, central to the human ability to develop novel thoughts. Recording is important, but it is only part of the story, and certainly not the most interesting one: By providing us with a way to record ideas and emotions, writing freed us from the memorization that oral tradition required and thus prompted us to come up with new thoughts. Think of the effort made by the Greeks, who learned by heart whole passages of Homer's *Iliad* and *Odyssey.* The exercise did not leave them much time and energy to analyze and discuss the content of these literary works nor to refine them (apart from the changes caused by imprecise memory). Imagine if we had to memorize all of Plato's *Dialogues* instead of having the ability to read them as many times as we wanted in order to analyze, interpret, and understand their meanings and so develop our own theories about truth, justice, or beauty?

Literacy studies (Bazerman, 1988; Goody, 1977; Goody & Watt, 1963; Havelock, 1963; Ong, 1982/2002) show how writing represented a radical shift in human history, impacting various cultural systems as well as human cognitive abilities. Thus, Goody (1977) shows how the transition from magic to science or from the "savage" mind to domesticated thought (Lévi-Strauss, 1962) corresponded to the shift from oral communication to the development and use of writing. Writing also influenced the organization of society. For example, in writing-based societies, the judge referred to the written law, while the judge in oral societies grounded his decision on a set of relevant proverbs articulated in a meaningful way (Ong, 1982/2002). More specifically, the Greeks discovered how to define and express legal problems, as the use of writing allowed them to isolate abstract concepts (Martin, 1994). This capacity to abstract is described as "one aspect of the mental equipment of homo scribens" (Martin, 1994, p. 79).

Also, writing has had a special influence on the accretion of scientific knowledge, as it allowed people to develop knowledge that they could share and build on across centuries (Goody, 1987; Havelock, 1963; Ong, 1982/2002). Formal logic was an invention of Greek culture after it had internalized the technology of alphabetic writing and "made a permanent part of its noetic resources the kind of thinking that alphabetic writing made possible" (Ong, 1982/2002, p. 52). In contrast, because orality is situational rather than abstract, it does not feature elaborate analytical categories for structuring knowledge. Thus, oral cultures did not demonstrate formal logic (e.g., syllogism) because syllogism was intrinsically linked to writing (Goody, 1987).

The effects of writing go beyond the development of human societies and intellectual systems, such as philosophy, mathematics, literature, or logic. More profoundly, the act of writing not only mediates but also changes thoughts themselves (Olson, 1977; Vygotsky, 1962), supporting the creation of abstract and novel ideas (Wolf, 2008). As Maryanne Wolf noted, more than the first alphabet, the act of writing itself had a major impact in promoting "the development of human thought in human history" (2008, p. 65). Indeed, as children learn to read other people's thoughts and write their own, they engage in a powerful and productive relationship between written language and new concepts. Wolf further argued that the introduction of the Greek alphabet increased cognitive efficiency (indeed, because of the economical use of characters, alphabetical systems reduce the time and attention needed for quick recognition). This increase allowed more people to develop novel thoughts at a younger age.

In her wonderful studies on the development of reading, Wolf described how, as children learn to decode the alphabet and become fluent readers, the world of reading and the knowledge it provides opens to them. They also learn how to think in a reflective and critical way. That is, fluency is not about reading fast but rather about reading fast enough that the reader has time to think and understand. As a reader is freed from the effort of deciphering written words, he or she is able to reflect on their meaning, feel emotions, make hypotheses, and develop new ideas.[1] Wolf showed it is not that fluency provides better comprehension but rather that a reader who is fluent gains an important resource: time to think. Fluency marks the moment when the young reader, going "beyond the information given" (Wolf, 2008, p. 132), has the ability to predict not only what the text says but also what it doesn't say. That time to think is also provided by the writing brain, which is intrinsically connected to the reading brain, and which allows writers to focus on the encoding of their emotions and thoughts. That time to think is also what makes writing such a creative mode and what has allowed humans over centuries to develop literature, mathematics, and philosophy. This process is not innate, argued Wolf, but it is repeated by each child who learns how to read and write and provides access to an abstract and reflective type of thinking.

Once we acknowledge the crucial role writing plays in enabling and shaping our thinking capacities, as well as in organizing society (e.g., facilitating literate societies and their logical, mathematical, and philosophical systems), we are still left with several questions: How does writing support knowledge development, not only at the societal level but also in small groups and in organizations, and what are the specific writing mechanisms that support knowledge development? Indeed, while literacy scholars refer to some of writing's dimensions, such as its fixedness or analytical capabilities, they do not focus on how these dimensions support the development of new theories or of the organizational knowledge that is crucial to the coordination of complex activities.

Moreover, historical and psychological studies in the literacy tradition do not address two other major functions of writing: the expression of emotions that it permits or the creation of bonds among geographically dispersed individuals. Indeed, the development of writing supported the growth of communities since people (occasionally, outsiders, such as women) could exchange ideas and emotions with kindred souls. These written exchanges further established their identity as community members. In this chapter, we identify more precisely the ways in which writing achieves its power in the three areas of expressing emotions, developing knowledge, and building communities. Before describing the three powers of writing, we introduce four mechanisms that enable those functions: objectifying, addressing, reflecting, and specifying.

THE FOUR MECHANISMS OF WRITING

Inspired by studies of writing in the literacy tradition and their provocative hypotheses regarding the creative nature of writing, we chose to examine the process the writer engages in during the writing act and, to a lesser extent, the process in which the reader engages. Approaching writing in this way allowed us to understand how (through which mechanisms) writing generated its powers. By analyzing several correspondences (among employees of distributed organizations, scientists engaged in distance collaborations, and writers and their family and friends) we identified four mechanisms—objectifying, addressing, reflecting, and specifying—that help explain writing's three main powers: It enables emotional expression, it fosters knowledge creation, and it allows participation in the public sphere. Our analysis suggests that objectifying and addressing are the two main mechanisms of writing, and that, in turn, they afford two other mechanisms, reflecting and specifying, which are central to writing's formidable capabilities.

Objectifying

Who hasn't pondered a problem and figured it out, only to find later, while writing it down, that the formerly crystalline notion is vague, unclear, and fleeting? Only after scribbling and drafting can we eventually articulate the idea and share it with others. In fact, it is through the act of *writing it down* that we have been able to *objectify* it. Hence, we can say that one thinks "through writing": The process of writing helps to clarify our initially amorphous ideas and feelings. Through objectifying, an idea or emotion can be expressed clearly and can be reflected upon, specified, and modified by the writer as well as others. Therefore, the fixed nature of the writing product—the written text, so often criticized, especially by Plato in his description of written texts as "dead signs," "that preserve a solemn silence if one questions them" (*The Phaedrus,* 275d [Plato 1940a])—has, in fact, positive implications, as it forces writers to clarify and articulate an idea for themselves and for others. In fact, historians like Martin (1994), in his book *The History and Power of Writing,* noted how "recourse to graphic expressions seems to have represented man's need to give visual form his interpretations of the external world; to fix those interpretations and make them concrete in order to define them better; to take possession of them, communicate with the superior forces, and transit what we have learned to his fellows" (Martin, 1994, p. 4).

We often think of writing in terms of written texts and documents, such as the written tablets or papyrus found in museums, the files of archived correspondences generated by big companies like the Hudson's Bay Company and East India Company, both described as the ancestors of today's distributed organizations, or the scribbled notes of scholars, friends, or lovers that have been archived and that we can now study. However, these traces and materializations are only one part of writing; they are the result of the objectifying process. These outputs are important as they allow the sharing of knowledge among people across centuries. But these outputs are only one facet of writing. The other facet is the writing process itself, the activity of producing these texts or documents—the act of objectifying feelings, thoughts, impressions, and ideas in the act of writing. By producing a written text, we share emotions or ideas with others. Indeed, in the act of writing, objectifying provides the opportunity for the writer to articulate a best practice, a rationale for action, or a complex argument or emotion. Sometimes, just by writing down a technical problem, we can clarify the issue and sometimes even solve it, or at least start collaborating with someone who can help. Moreover, the written text can always be modified. Letters, e-mail, and presentations can be revised and revised again; even books that result from multiple drafts can be revised with the release of new editions. Similarly, by writing down emotions such as anger or frustration, we can clarify their underlying causes and then explain our position to the person responsible for engendering those feelings. In the example of the soldier and his wife in Chapter 1, for instance, abandoning Skype to write letters allowed them to better articulate their emotions, negative and positive, and rebuild their relationship.

Addressing the Reader

Whether we write a letter, speech, or book, we always write *to someone* and try to adapt our text to the (potential) reader's perceived needs, knowledge level, attitudes, and interests. When we speak to someone, we also address that person, but we can see, or at least hear the reactions of our interlocutor and thus react and adapt what we say on the spot. On the contrary, when we write, we cannot see the reader, and the reader cannot ask questions and request complementary information. Therefore, we imagine the reader and provide contextual information but also guess the reader's questions or interpretations. This leads us to make decisions about content, wording, and tone. We draft e-mail or letters to find the right tone or wording, taking care to ensure that our message will be understood properly and not misinterpreted. Because writing is asynchronous, the addressing process can take as much time as is needed until the written text captures our message fully and accurately. In contrast, in face-to-face situations, we need to be constantly on our toes, constantly reacting to small changes in the partner's words or nonverbal behavior. Again, if we consider the example of the soldier and his wife, the addressing mechanism afforded by writing was fundamental in allowing them to express their feelings in a way that was mindful of the partner's full range of possible emotions and reactions.

Reflecting

Because writing allows us to objectify ideas and write them down (on a piece of paper, Post-It note, e-mail, or Word document), it also allows the writer as well as the recipient to read those ideas several times and to reflect on them. Because it is asynchronous, writing allows the writer to consider ideas or emotions before, while, and even after writing them, as the writer can always modify a draft or start a new version. The reader also has the opportunity to reread the message to ensure understanding and to reflect on it. In the example discussed in Chapter 1, the soldier and his wife kept each other's letters, thought about them before replying, and reread them regularly to revive their emotions and reinforce their relationship. Therefore, writing and its counterpart, reading, are reflective activities that allow people (if they choose) to deepen their thoughts and become aware of new ideas and feelings. Our use of the term *reflecting* broadly encompasses rational, conscious reflection and deeper awareness that may take place at the unconscious level. Historically, the development of reflective thought by *Homo sapiens* has been associated with the ability "to develop both graphic schematization and verbal conceptualization as he attempted to analyze the universe" (Martin, 1994, p. 6). It is interesting to note that while objectifying—the writing down of ideas and the trace it leaves—supports the reflecting process, the reverse is also true: Because of their reflecting abilities, humans are able to objectify new and existing thoughts and feelings—first through graphics and later through writing.

Specifying

As we try to articulate our ideas and emotions, we usually attempt to be as specific as possible, adding details and using precise language. Moreover, while addressing the reader, we make hypotheses about his or her understanding (and possible misunderstanding) as well as potential questions and interpretations, and we strive to be as clear and detailed as possible so that the reader can easily follow our intention.[2] Oral communication unfolds in time, allowing us to go back and forth and to sometimes incompletely formulate our thoughts without being too disruptive to the communication process. Indeed, audio or video recordings of conversations highlight how unfinished and diluted an oral conversation can seem when listened to out of context. Writing, which takes place on a two-dimensional surface, taking the "flux of words unfolding in time and transcrib[ing] them along a line in space" (Martin, 1994, p. 8), forces us to be as specific as possible. Because correspondents often live in different contexts, the effort to be understood on the basis of the written word also imposes the need to be specific and precise. (This, of course, does not mean that writing cannot be ambiguous; some people master the art of ambiguous writing.) Moreover, writing is always interpreted and can therefore be understood in various ways. However, because of its fixed and asynchronous nature, writing tends to encourage the writer to be more specific. For example, the wife and husband in Chapter 1 enjoyed writing to each other because they could impart details and rich descriptions of their everyday life as well as their thoughts and feelings.

Although we present the four mechanisms of writing (see Table 2.1 for an overview) as analytically independent, they should be understood as interdependent—overlapping and supporting each other. As noted above, objectifying and addressing are the two main mechanisms, and both trigger another mechanism. While objectifying mostly enables reflecting, it also indirectly enables specifying: For example, as the writer clarifies her thinking while writing down, she may enhance her ability to further specify her ideas and emotions. Similarly, while addressing supports the specifying mechanism, it also indirectly enables reflecting: Making hypotheses about the reader's possible interpretations and questions may push the writer to deepen his reflection on the written text.

The four mechanisms constitute an analytical tool that allows us to explain three main processes that are important for the effective functioning of formal and informal organizations: emotional expression, knowledge development, and community building. Although these mechanisms are

TABLE 2.1 The mechanisms of writing

Writing mechanisms	Definition
Objectifying	The act of expressing, writing down, and externalizing an emotion or an idea, the result of which can be reread, reflected upon, modified, and shared with others.
Addressing the reader	The writer always writes *to someone* and adapts the text to that person's perceived needs, knowledge level, and interest.
Reflecting	Once an emotion or an idea is written, both the writer and reader can read it several times and reflect upon it.
Specifying	The act of being precise and giving details as the writer tries to articulate clearly his or her points and answer the reader's possible questions.

based on analyses of letters, they are intrinsic to writing in general. The identification and examination of mechanisms is at the core of social science advancement (Hedstrom & Swedberg, 1998; Merton, 1968). Thus, the mechanisms of writing we investigate in this chapter are critical to revitalizing scholars' conception of writing as a creative modality, and they provide a constructive way of reconciling conflicting claims and findings regarding the capabilities of written forms of communication.

THE THREE POWERS OF WRITING

Writing has been criticized since its invention, and today it is still considered an impoverished mode of communication by many organizational communication theories, such as media richness theory, that ignore the historical role writing has had for organizations, even as written documents, such as letters, reports, and memos, have played a central role in the development of modern organizations (O'Leary, Orlikowski, & Yates, 2002; Yates, 1989). (For an overview of the role of the memo in the emergence of modern organizations, see the sidebar The Memo's Role in the Emergence of Modern Organizations.) Such theories also ignore the fact that, despite the novelty of the medium, online written interactions

THE MEMO'S ROLE IN THE EMERGENCE OF MODERN ORGANIZATIONS

Written documents such as letters, reports, and memos have played a central role in organizations (O'Leary et al., 2002; Weber, 1968; Yates, 1989). As Weber observed in *Economy and Society,* "The management of the modern office is based upon written documents (the 'files')" (Weber, 1968, vol. 2, p. 957) and writing is necessarily connected to the very idea of the office or "bureau" as the spatial means of organizing scribal labor. Furthermore, scholars directly focusing on the role of writing in organizational communication have uncovered this mode's crucial contribution to the development of the modern organizations. Most prominently, Yates (1989), in *Control Through Communication* and several related articles, describes the emergence of the memorandum as a genre of writing that corresponds to the advent of large corporations in the late 19th century and the development of new management techniques. The term *memo* was first used in the late 1870s and early 1880s, becoming common in the 1920s. As Yates shows, the memo transmits information, gives direction, and makes recommendations, all crucial activities in the organizing of work. Before the late 19th century, although businesses consisted mostly of small entrepreneurs, letters were still the primary form of written communication, but as organizations grew and new kinds of managerial practices emerged, internal communications across different offices and different levels of management called for another type of communication. The memo as a new genre of internal communication emerged, and it contributed to the effective management of large, complex, and increasingly distributed organizations.

in contemporary organizations rely on a basic and ancient mode of communication: writing.

Theories such as the media richness theory and the social presence theory tend to emphasize two features of writing that, in their view, make it an impoverished mode of communication: fixedness and asynchronicity. Because of these dimensions, writing can only support the recording and sharing of simple, explicit knowledge (such as lists of numbers, rules, etc.), but it cannot support the creation of knowledge or the sharing of complex, ambiguous ideas (Daft & Lengel, 1984, 1986). Nor can we

CONTEMPORARY DEBATES ABOUT WRITING'S POWER

In organizational theory, the trend toward criticizing writing's capabilities has taken a peculiar form. Instead of focusing on writing as a fundamental mode of communication, the criticism focuses on the particular medium in which the writing occurs. Thus, the invention and extensive use of communication media in organizations was accompanied by a set of theories that tried to match the task to the medium.

The media communication theories tend to consider written communication a poor means for expressing complex knowledge and emotions. Thus, media richness theory (Daft & Lengel, 1984, 1986), the most prominent theory of communication media, indicates that the different media vary in terms of their "richness." It holds that communication media have different capacities for resolving ambiguity, negotiating varying interpretations, and facilitating understanding. Using four criteria—the medium's capacity to provide immediate feedback, the number of cues and channels available, personalization (the degree to which it focuses on the recipient), and language variety (Daft, Lengel, & Trevino, 1987; Daft & Wiginton, 1979)—the theory posits the different media on a continuum extending from rich to impoverished communication. The "richest" medium is face-to-face interaction, and the other media are classified in decreasing order: telephone, personalized letters, memos and e-mail, impersonal written documents, and numeric documents. This conceptualization leads Daft and Lengel (1984, 1986) to claim that for difficult, equivocal topics, managers (should) speak face to face with their audience and reserve memos, bulletins, and reports—and e-mail, although it is only mentioned in later studies—for topics that are better understood and specific. The assumption is that only face-to-face communication can provide a wide array of cues and language usages that enable exchange partners to express themselves fully and form a sense of one another.

Consequently, studies informed by these theories tend to associate the limitations of computer-mediated communication with the limitations of the written word, whose only positive attribute seems to be its recording capability. It is implicitly suggested that for the idea to be effective, computer-mediated communication needs to be supplemented and augmented by video or multimedia technologies that bring it closer to the ideal of face-to-face exchange (Daft et al., 1987). Even scholars who have suggested that computer-mediated communication can have positive characteristics, taking into account elements such as context or experience (if people know each other well, if they master the technology well) consider face-to-face communication to be ideal. For example, Walther and Burgoon (1992) claim that if more time is given, computer-mediated groups can develop similar or even better relationships than those formed in face-to-face settings. The underlying assumption here, as in all the criticisms of writing, is that most people believe face-to-face communication (the "live moment" of the spoken word) is the ideal and richest type of communication.

express emotions or build communities through writing because we miss the richness of face-to-face interactions that provide us a sense of "social presence" and direct feedback about others' reactions (Short, Williams, & Christie, 1976). Empirical studies that rely on theories, such as media richness theory, constantly compare evolving media features and cannot explain the capabilities of writing to build relationships, foster community development, or create and share complex knowledge. (For a discussion of contemporary debates on writing, see the sidebar Contemporary Debates About Writing's Power.)

Such studies are unable to illuminate a phenomenon, such as the Hudson's Bay Company, one of the first distributed companies, or the development of scientific theories by communities of researchers who spanned many countries and who communicated mostly through correspondences (Collins, 1998). However, as we outline below, the four writing mechanisms we identified help explain such phenomena and provide a theoretical understanding of writing's capability to support the expression of emotions, the development of knowledge, and the building of communities.

Expressing Emotions

The four mechanisms are important when it comes to expressing emotions in organizational contexts as well as in personal instances, such as Virginia Woolf's correspondence with her friends, husband, and family. Objectifying is central for expressing emotions because it allows the writer to articulate, externalize, and reflect on them, and thus, sometimes, make them more manageable. Addressing the reader in the emotional context is also crucial because it allows the writer to engage in a dialogue with the reader. Writers have to specify their nuances so that readers will understand them. At the same time, in this process, the writer often arrives at a deeper understanding of his or her own emotions. Emotional expression is not only in the realm of personal correspondences, though. Positive and negative emotions are also plentiful in organizational e-mail or in online forums. Think of "escalation" e-mails, in which people get angry at each other and the tone becomes more aggressive, and, typically, the number of other people copied multiplies. Oftentimes, people who engage in these written arguments wouldn't have them face to face, but the absence of the other allows them to express negative, or strong and personal, emotions. At the same time, some online forums relating to adoptions or to diseases, such as cancer, are incredible examples of how a whole array of complex positive and negative emotions can be articulated through writing.

Developing Knowledge

Scientists, such as Albert Einstein and Élie Cartan, who lived in different countries, or members of a distributed organization, such as the Hudson's Bay Company, could create and share knowledge while writing. The four mechanisms allowed them to articulate ideas, share them, and develop them collaboratively. For example, through writing, and particularly through objectifying and specifying, Einstein was able to articulate his ideas in a way orally he could not because of their complexity.

Furthermore, Einstein addressed his writing to Cartan—providing him with context and explanations about his rationale for developing one particular assumption rather than another. He could then share them with Cartan, who, in turn, could reflect on them and develop them further, engaging in a collaborative process with Einstein. Similarly, contemporary scientists share results, improve experiments, and write academic papers via e-mail. There are also collective efforts, such as open source software development, where developers create and develop software code through online platforms; online forums, where communities of people with common interests and expertise share knowledge and practices; and, more recently, open innovation platforms, where people solve challenges and find solutions and new ideas about which they write and post. These efforts are possible because writing allows them, through objectifying, not only to write *down* ideas that can be shared and developed but also to think and develop ideas for themselves through the process of writing.

Building Communities

Addressing the reader, and the dialogue it enables, are also critical to community development, as is reflecting, which permits community members to develop a better understanding of both the ideas and emotions they have in common. As they write about their shared beliefs and sense of identity, community members objectify and perform their shared identity. At the same time, specifying emotions and ideas also leads to the development of a sense of *we*-ness among people. Many online forums do not evolve but instead remain as online bulletin boards, with a few people joining to post a question or a short reply, but some do grow as communities would, with members feeling they are involved in a group with similar values and a common identity. Often the transformation of forums into communities rests on the expression of emotions and the development of knowledge through (written) posts. The three powers of writing are somewhat related because they originate from the four mechanisms. Thus, engaging in knowledge development and exchange, and expressing strong trust in their correspondents enhance their community bonds; at the same time, belonging to the same community supports deep and genuine exchanges.

The three chapters that follow discuss each of these powers and demonstrate how the mechanisms of writing enable the expression of emotions, the development of knowledge, and the building of communities. We achieve this by analyzing several sets of historical correspondences of writers, scientists, and distributed organizations.

METHODOLOGICAL APPROACH

There has been little theoretical guidance for identifying the specific ways in which writing supports the three types of accomplishments of interest to us, so we took an inductive approach. Most of the debates discussed above forget that many of the communication technologies we currently use are writing based, and that we have been writing for millennia, exchanging ideas and emotions, and developing communities. Thus, we decided to investigate how writing supports these processes by taking a historical perspective and focusing on the mode and avoiding the discussion about the influence of the media. To do so, we conducted a discourse analysis of several sets of correspondences in organizational contexts (e.g., letters sent by commanders of the Hudson's Bay Company to its headquarters in London), in scientific communities (Marin Mersenne's network, Madame du Châtelet's correspondence, and the Einstein–Cartan correspondences), as well as the letters of famous writers, such as Virginia Woolf with her friends and family. Then, in the second part of the book, we show how the mechanisms of writing that we uncovered are relevant for analyzing and understanding current online communicative practices.

Our interest in letter writing is first an interest in writing, letters being a specific example of writing. Letters as a genre allow us to focus on the dialogue between the reader and the recipient, and by their dialogic nature, letters are the ancestors of e-mail communication, which plays a key role in today's organizations (Byron, 2008). Also, the early versions of media richness theory were based on letter writing instead of e-mail (Daft & Lengel, 1984). While there are important differences between letters and e-mails, they are both malleable genres ranging from short, spur-of-the-moment notes to more elaborate and sometimes quite well thought out and elaborate correspondences (Orlikowski & Yates, 1994). Moreover, a study by Oliveira and Barabási (2005) showed an interesting similarity in the patterns of letter management and e-mail management. Analyzing Einstein's and Darwin's letters in terms of writing patterns (particularly the intervals between their receipt of a letter and their responses), Oliveira and Barabási (2005) found that the delay followed a "power law," much like e-mail does today. The authors point out that the similarity in scaling laws between electronic and written correspondence points to a "fundamental pattern in human dynamics" (p. 1251).

It is important to note that our choice has been motivated by a theoretical concern to separate the affordances of the various communication media from the capabilities of writing. Because we now live an era of media and channel profusion, focusing on a time when face-to-face interaction and letter writing were the only communication options available allowed us to better understand the full potential of writing. The emphasis on letters, as a writing genre, also gave us a better understanding of the power of writing, not only as a supportive means of recording what already exists (as exemplified by memos and accounting books) but also as a supportive means of creativity—the production of knowledge and interpersonal bonds and the expression of emerging emotions. Moreover, we explain how letters, as a genre, are situated at the intersection of the public and private realms. This boundary position makes them relevant for understanding current communicative practices that raise issues about publicity and privacy.

The choice of the particular empirical contexts—letters of the Hudson's Bay Company and Virginia Woolf and the posts on the open source software development and OpenIDEO forums—was also guided by our theoretical focus: the powers of writing. Since the powers of writing are most prominent in the case of successful distributed organizations relying almost exclusively on the written mode, and in that of novelists who are masters of the written word, studying such cases maximizes the chances of observing the powers of writing as well as the mechanisms on which these powers rely.

Our approach was qualitative, inductive, and iterative while defining the four mechanisms of writing. First, we both read the chosen correspondences using a discourse analysis approach (Phillips & Hardy, 2002). The first important realization occurred when we uncovered the importance of the objectifying and addressing processes at play in letters. During this first phase, we realized that writing seemed to possess specific mechanisms that were central to the expression of emotions, the development of knowledge, and the building of communities. In the second phase, we familiarized ourselves with the fragmented body of research represented by literacy theories. Armed with our knowledge of literacy theories as well as with the insights from the first round of analyses, in a third phase, we reread the correspondences, remaining open to new patterns emerging from the data. In this third stage we developed theoretical categories by comparing and contrasting examples from the correspondences in analyzing the data (Glaser & Strauss, 1967). During this third phase, we identified categories and then grouped them into themes, which we called mechanisms because they help explain important processes for the three powers of writing. In the fourth phase we examined the specific practices supported by writing for each power and in each setting.

CHAPTER

3

Expressing Emotions Through Writing

> Thank you for writing to me often, the one way in which you can make your presence felt, for I never have a letter from you without the immediate feeling that we are together.
>
> —Heloise's first letter to Abelard, *Letters of Abelard and Heloise* (1974)

How can one express something as changing, complex, and subtle as emotions or develop relationships using such a "poor" mode of communication as writing, which is devoid of nonverbal cues and lacks the warmth of face-to-face interactions? In this chapter, we illustrate how expressing emotions through writing is neither impossible nor a tour de force. We examine the reasons for which writing has often been seen as a limited mode for expressing emotions and show how, in fact, some of these so-called limitations allow the expression of emotions. More specifically, we discuss how the four mechanisms of writing are central to expressing emotions in two different contexts: organizational letters from the Hudson's Bay Company and personal letters between Virginia Woolf and different members of her community.

Notwithstanding the view that writing is poor at emotional expression, the reality of correspondences—personal and organizational—suggests a different view of writing. Since the earliest preserved correspondences, there are numerous examples of emotions expressed and shared in writing. Letters exchanged in ancient Rome testify to the complexity of feelings expressed by their authors. Indeed, the letters of Emperor Trajan (AD 53–117) to his friend Pliny the Younger, and those of Pliny (AD 61/63–113) to his wife, Calpurnia, are filled with nuanced descriptions of emotions and attest that letters were used to build or maintain relationships with people whom the writers rarely saw (De Pretis, 2003).

Examples of famous, as well as less famous, correspondences in which writers shared thoughts and emotions with their friends or relatives, or even unknown others, abound. Several famous literary couples—Heloise and Abelard, George Sand and Alfred de Musset, Simone de Beauvoir and Jean-Paul Sartre—have developed and deepened their relationships through writing (Poisson, 2002). More than just deepening a relationship, writing has been in several cases the starting point of the relationship, as it was for Elizabeth Barrett and Robert Browning. This was also the case for many anonymous young soldiers and their "war godmothers" during the First World War. In the spring of 1915, the French government created and promoted *marraines de guerre* (war godmothers). These *marraines de guerre* were supposed to "adopt" a soldier without a family and write to him regularly, providing personal support (Darrow, 2000). Originally, the aim of the *marraines de guerre* was patriotic—to give courage to the soldiers at the front and to personify France for which they were fighting. Some soldiers explained how writing to someone, and knowing that someone was reading and replying to their letters, made them feel stronger in the battle but also reassured them that if they were to die, someone would mourn them. Quickly, the style of the letters shifted, and many letters became more sentimental than patriotic and flirtatious relationships emerged. In 1918 and 1919, marriages between the *marraines de guerre* and their *filleuls* (godsons) took place (Le Naour, 2008).

Emotions are also central in organizations, which can be described as "emotional arenas" (Fineman, 1993, p. 9). If we acknowledge that the experience of work is saturated with emotions (Ashforth & Humphrey, 1995), one can only recognize their impact on communication in distributed contexts, such as the Hudson's Bay Company or the East India Company, where the managers at the headquarters have often never met the employees in the various posts and have no idea of the hardship of their lifestyle. A variety of emotions, while less directly expressed than in personal letters, infused even such bureaucratic documents as the letters exchanged among company employees. In this chapter, we show how writing's mechanisms enable people to effectively express emotions, negative or positive.

THE WRITING MECHANISMS AND THE EXPRESSION OF EMOTIONS

Because it is evanescent, flexible, "live," and involves nonverbal expressions, oral expression is generally believed to be more powerful than writing to express the richness and ever-changing nature of one's feelings. As the French philosopher Jean-Jacques Rousseau wrote "I have always said and felt, real joy cannot be described" (1763/1959, p. 98). In contrast, writing is perceived as having two main shortcomings. First, because it is "fixed" and asynchronous it does not allow for a real dialogue, which is crucial to expressing emotions. Indeed, dialogues allow people who are expressing their feelings to respond to the reactions of others, in particular, to nonverbal behaviors (e.g., tears, eye rolling, head shaking). In contrast to oral communication, written texts are often criticized as being "dead signs" (*The Phaedrus*, 227a–279b [Plato, 1940a]; Rousseau, 1763/1959). As we explained in the previous chapters, Plato, in the *Phaedrus* (227a–279b), described spoken language as being alive and written texts as being dead, the former being open to reinterpretation in face-to-face interactions while written texts, although open to interpretations, do not permit dialogue or questions and do not alter with changing contexts. Writing, which fixes thoughts and conversations in time, is viewed as antinomical to the ever-changing nature of emotions.

Second, writing is seen as being too universal, too general. Indeed, when writing about emotions such as anger, fear, or love, we are referring to *general* emotions, which must be able to describe any emotions, and which consequently lack specific details and nuances. Writing's capacity to generalize seems to prevent it from allowing us to express the individual specificity of our emotions. Rousseau (1781/1990) developed a similar argument for the expression of emotions and argued that one cannot express emotions through writing.

On the contrary, we argue that—paradoxical as it may seem—the objectifying mechanism is key in the expression of emotions. Indeed, feelings, which are often hard to express because they are fleeting and immaterial, become objects through the writing process that can be expressed, reflected upon, and shared. The objectifying process allows the writer to articulate and clarify his or her emotions. Indeed, research suggests that writing about traumatic life experiences provides an opportunity for increased insight, self-reflection, and organization of one's perspective of the problem as opposed to merely venting emotions (Smyth & Pennebaker, 1999). For example, a cancer patient wrote: "Writing about my cancer brought out feelings that I hadn't dealt with. My understanding of these feelings brought many, many conversations with my husband and close friends which in turn brought out things they were feeling and were afraid to say. I feel that all of this has brought us all a lot closer and opened a door wider for conversations" (quoted in Stanton & Danoff-Burg, 2002, p. 41).

As the quote indicates, writing down one's feelings allows the articulation of emotions not yet named, thus enabling a better understanding of one's state of mind. Because writing is a solitary exercise, lacking immediate feedback, it allows reflection, self-analysis, and greater understanding. Hence, this lack of immediate feedback, which is often referred to by studies criticizing online text communication (e.g., Byron, 2008; Daft & Lengel, 1986), is in fact a strength when it comes to expressing emotions. Indeed, thanks to the absence of immediate feedback, writers can reflect more deeply because they don't have to manage their nonverbal and facial expressions, as they have to in face-to-face interactions. In some cases, writing may facilitate the sharing of difficult emotions (such as anger and sadness) because the writer does not have to take into consideration the immediate reaction (perhaps negative) of the audience. In

TABLE 3.1 Writing's mechanisms and their effects on emotional expression

Writing mechanisms	Practices	Effects on expressing emotions
Objectifying	1. Describing feelings and emotions.	Helps articulate the emotions.
	2. Referring to previous letters.	Articulated emotions become an object that one can then manage.
	3. Using images.	The letter, kept and sometime read many times, becomes a symbol of the relationship.
Addressing the reader	1. Providing contextual information (e.g., on when the letter was written or received) or elements in the life of the writer that might explain better how he or she felt at the moment the event described took place, or when he or she wrote the letter.	Facilitates understanding as the recipient is provided with contextual information.
	2. Referring explicitly to the reaction of the reader (e.g., "You might wonder why I did not say it immediately…").	Imagining the reader's reactions helps the writer articulate emotions (i.e., putting emotions into words).
	3. Providing extra information to clarify an emotional reaction that might be misunderstood (e.g., "If I did not reply immediately, it is because I was so moved by what happened…").	
Reflecting	1. Referring to the time taken to write or read a letter and to the interpretative process (e.g., "I received your letter a few days ago, but I needed some time before replying…"; "I was so moved when I received your letter that I needed some time to respond…").	Facilitates and enhances the writer's understanding of his or her own feelings and emotions.
	2. Referring to the reading of the letter and questioning it (e.g., "As I was reading your letter, I started wondering if maybe your frustration was not due to a miscommunication…").	Helps create a dialogue with the reader and support the development of a relationship.
	3. Referring to multiple readings of the letter.	

Specifying	1. Long descriptions, use of many adjectives.	Allows writers to articulate their emotions without feeling that they oversimplify them.
	2. Using complex sentences that sometimes can reflect the complexity of emotions (such as in stream of consciousness).	Allows writers to have a sense of empowerment rather than being overwhelmed.
	3. Providing a lot of details and/ or nuance.	

fact, writing emotions down might be positive. For example, people who wrote down their traumatic life experiences experienced both psychological and physical health benefits (Lepore & Smyth, 2002; Pennebaker, 1990; Smyth & Pennebaker, 1999; Stanton & Danoff-Burg, 2002). These studies illustrate how writing, by objectifying emotions, frees up the mind to allow catharsis and the expression of unarticulated and confused emotions. Emotions—which are often difficult to share or to analyze for oneself because they are internal, ambiguous, and changing—through writing become an object that can be analyzed and shared.

The quote from the cancer patient also suggests the need for nuance in the second main criticism of writing, namely, its universal nature (Auroux, Deschamps, & Kouloughli, 1996), which does not allow it to support the expression of emotions that are perceived as individual. As the cancer patient explained, through the objectifying and reflecting mechanisms involved in the writing process she arrived at a deeper, more specific, and nuanced understanding of her feelings. Even though she did not explicitly state it, a deeper understanding of her complex feelings was also facilitated by her writing about them *to someone* (i.e., by the addressing-the-reader mechanism). Social-cue theorists have noticed that oral expression is marked by both the individual (gender, health) and social (nationality, social status) characteristics of the speaker (Short et al., 1976; Sproull & Kiesler, 1986). Yet, addressing our writing to someone affords the enactment of our individuality. Furthermore, writing can become a vehicle for the immediate expression of our spontaneous leaps of emotion, the instrument capable of translating the fluctuations, the incoherence, and the contradictions of passion and emotion (Rousset, 1966, p. 77), an aspect, which, again, strengthens the possibility of a true dialogue. Table 3.1 shows the relationship between each of the mechanisms of writing and the expression of emotions.

In the following sections we will illustrate how writing, through the enactment of its four mechanisms, allows the expression of emotions, both in an organizational context (the letters exchanged among the managers of Hudson's Bay Company based in the headquarters in London and their commanders based in various posts in Canada) and in a personal context (the correspondence of Virginia Woolf with her friends and family).

EXPRESSING EMOTIONS IN ORGANIZATIONAL LETTERS

Organizational letters (i.e., documents produced by members of formal or informal organizations and focused on their work and the functioning of their organization) may seem to be dry, administrative documents, devoid of emotional expression. However, this is not the case, as our examination of the letters exchanged by employees of the Hudson's Bay Company, an organization whose operations spanned two continents, shows. These letters illustrate how organizational life is filled with emotions (Ashforth & Humphrey, 1995).

The Case of the Hudson's Bay Company (1703–1740)

The Hudson's Bay Company was a trading company that supplied basic goods such as food, tools, guns, traps, clothing, and shoes to settlers in remote areas along the Hudson River in North America. It is the oldest company in North America. It was incorporated on May 2, 1670, with a Royal Charter from Charles II. Its headquarters were in London, England, but most of its operations are in North America. The company established the first permanent English settlements on the Canadian mainland by planting remote and almost monastic posts on the shores of the Hudson River. Before countries existed on the North American continent, the company was the effective ruler of an immense territory: In 1870 when its monopoly was abolished, the company controlled a territory of almost 4 million square kilometers. The posts consisted of little groups of white men who carried out those tasks upon which their masters' trade and their survival depended. In this large, successful, long-lasting organization, letters were almost the only means of communication among distant members (O'Leary, Orlikowski, & Yates, 2002).

The Hudson's Bay Company kept detailed records of all its activities. The London governor, deputy governor, and committee members made decisions about the pursuit of trade but rarely visited the North American continent where the trade occurred. The only contact of the posts with the outside world were the ships that sailed each year from the River Thames in May and anchored off Albany or York Fort in August (if they were lucky). Their stay was brief. The ship uploaded its outward-bound cargo of trade goods and supplies and received in exchange the homebound cargo of fur. Then it had to sail fast to avoid being caught in the ice. It bore home each year's general letter addressed to the company's London committee from the post to which it had been sent. These letters were generally four to six pages long (but were up to 10 pages), and they often had an inventory attached to them. The letters contained answers to the instructions or questions that the ships yearly brought from London and each post's story of the problems it was facing. Along with the

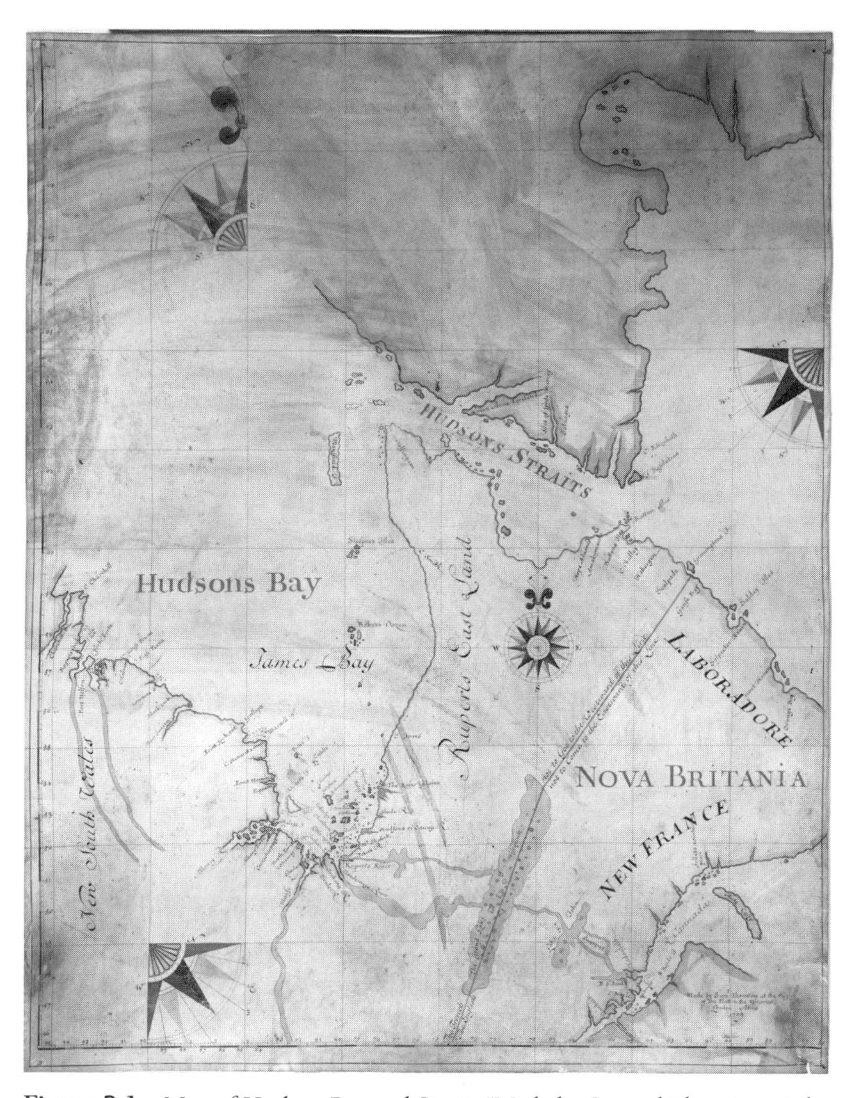

Figure 3.1 Map of Hudson Bay and Straits "Made by Samuel Thornton at the Signe of the Platt in the Minories London Anno 1709." By Samuel Thornton. HBCA G.2/1 (N4605).

Credit: Hudson's Bay Company Archives, Archives of Manitoba

annual letter, the post commanders also sent a fort journal (similar to a ship's captain's log). These were almost the only sources of information for the headquarters. All these documents had to be written in duplicate, one copy for the post and one for London. The first of the regular series,

Figure 3.2 *York Factory* (1770s?), by Samuel Hearne. Engraving, colored, March 1, 1797. Engraver: Wise. HBCA P-228 (N5411).

Credit: Hudson's Bay Company Archives, Archives of Manitoba

kept by the company in their archives, is an Albany letter dated1703. There were also country correspondences between the posts. After 1740, the country correspondence had to be prepared in triplicate: one for the recipient, one for the originator, and one for London. The collection of letters became a source of information that would allow the headquarters and other posts to know about the inventories and activities of the posts but also to understand the rationale for decisions and actions, which could be referred to and used—as best practices—by other posts and commanders. We analyzed the 79 letters sent from the forts to the headquarters office in London from 1703 to 1740 ('Letters from Hudson Bay 1703–1740, 1965').

Our analysis of the Hudson's Bay Company's letters reveals a correspondence rife with nuanced and complex emotions. Perhaps, not surprisingly, the post commanders often expressed negative emotions—complaining about their life and work conditions and about supplies and equipment headquarters did not provide them—but they did so subtly so as to not upset managers in London or contest the hierarchy. Writing allowed post commanders not only to reply to the criticisms of the headquarters but also to express their bitterness and frustration.

Figure 3.3 *York Factory—Arrival of Hudson's Bay Company's Ship* [ca.1880]. Engraving (B&W & hand-colored) by Schell & Hogan. Taken from *Picturesque Canada* (Vol. 1, p. 317), 1882, Toronto. HBCA P-676 (N14407).

Credit: Hudson's Bay Company Archives, Archives of Manitoba

Yet, they also conveyed positive emotions, showing how motivated and involved they were in order to make headquarters trust them. More specifically, our analysis shows how writing allowed post commanders to express both negative and positive emotions through several practices: criticizing and complaining, justifying, and seeking empathy. Table 3.2 provides a summary of the writing practices and mechanisms as well as their effects on emotional expression we found in the correspondence we analyzed.

Expressing Negative Emotions: Criticizing and Complaining

Letters in distributed organizations such as the Hudson's Bay Company allowed both the managers at the headquarters in London and the employees in the various scattered posts to express negative emotions such as mistrust, frustration, anger, bitterness, and fear (of being lied to, for example). The managers in the headquarters had often not met the post

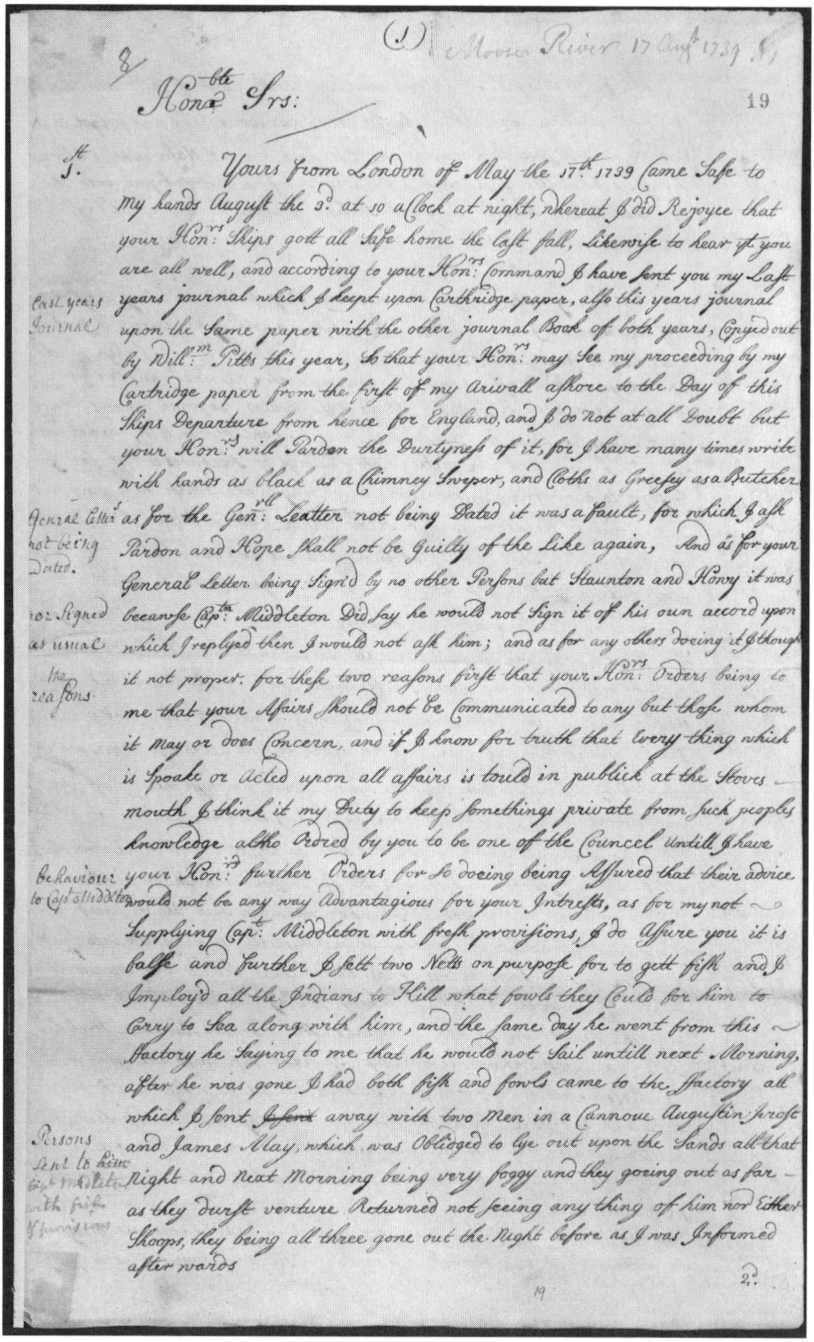

(1) Moose River 17 Augt 1739.

Honble Srs:

19

1st

Yours from London of May the 17th 1739 Came Safe to my hands August the 3d at 10 a Clock at night, whereat I did Rejoyce that your Honrs Ships gott all Safe home the last fall, Likewise to hear yt you are all well, and according to your Honrs Command I have sent you my Last years journal which I keept upon Carthridge paper, also this years journal upon the Same paper with the other journal Book of both years, Copyed out by Willm Pitts this year, So that your Honrs may See my proceeding by my Cartridge paper from the first of my Arivall ashore to the Day of this Ships Departure from hence for England, and I do Not at all Doubt but your Honrs will Pardon the Durtyness of it, for I have many times write with hands as black as a Chimney Sweper, and Cloths as Greesey as a Butcher as for the Genrll Leatter not being Dated it was a fault, for which I ask Pardon and Hope Shall not be Guilty of the like again, And as for your General Letter being Sign'd by no other Persons but Staunton and Hovy it was because Captn Middleton Did say he would not Sign it of his own accord upon which I replyed then I would not ask him; and as for any others doeing it I thought it not proper: for these two reasons first that your Honrs Orders being to me that your Affairs should not be Communicated to any but those whom it may or does Concern, and if I know for truth that Every thing which is Spoake or Acted upon all affairs is tould in publick at the Stoves mouth I think it my Duty to keep somethings private from such peoples knowledge altho Ordred by you to be one of the Councel Untill I have your Honrs further Orders for So doeing being Assured that their advice would not be any way Advantagious for your Intrests, as for my not Supplying Capt Middleton with fresh provisions, I do Assure you it is false and further I sett two Netts on purpose for to gett fish and I Imployd all the Indians to Kill what fowls they Could for him to Carry to Sea along with him, and the same day he went from this Factory he Saying to me that he would not Sail untill next Morning, after he was gone I had both fish and fowls came to the Factory all which I sent away with two Men in a Cannoue Augustin Frost and James May, which was Obliged to lye out upon the Sands all that Night and Next Morning being very foggy and they goeing out as far as they durst venture Returned not seeing any thing of him nor Either Sloops, they being all three gone out the Night before as I was Informed after wards

2d.

Figure 3.4 Letter from Moose Factory to London, dated August 17, 1739. HBCA A.11/43 fo. 19 (N16898).

Credit: Hudson's Bay Company Archives, Archives of Manitoba

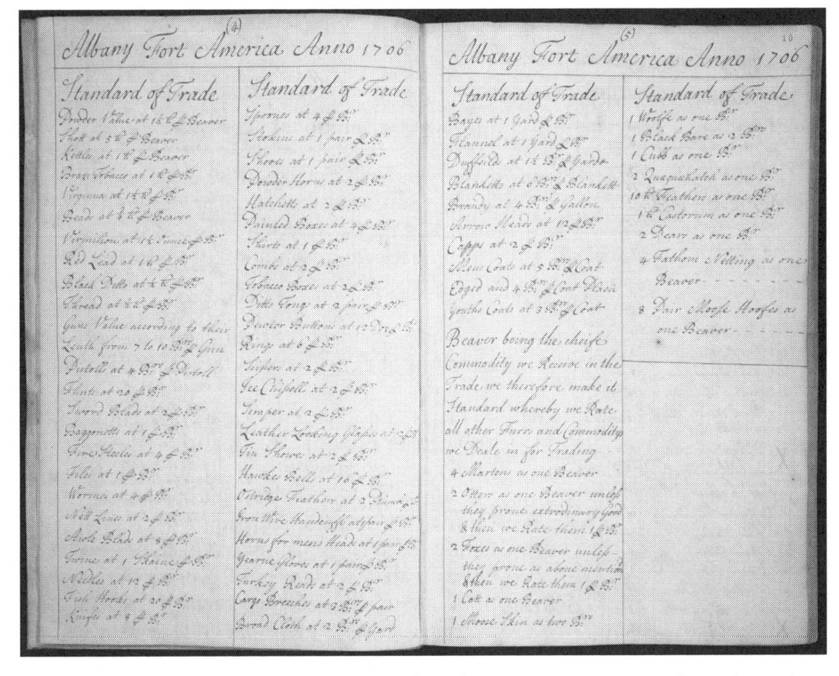

Albany Fort America Anno 1706

Standard of Trade

Powder Value at 1½ ℔ ℘ Beaver
Shot at 5 ℔ ℘ Beaver
Kettle at 1 ℔ ℘ Beaver
Brazel Tobacco at 1 ℔ ℘ Br.
Virginia at 1½ ℔ ℘ Br.
Beads at ½ ℔ ℘ Beaver
Vermilion at 1½ Ounce ℘ Br.
Red Lead at 1 ℔ ℘ Br.
Black Ditto at ½ ℔ ℘ Br.
Thread at ½ ℔ ℘ Br.
Guns Value according to their Lenth from 7 to 10 Brs. ℘ Gun
Pistolls at 4 Brs. ℘ Pistoll
Flints at 20 ℘ Br.
Sword Blades at 2 ℘ Br.
Baggonetts at 1 ℘ Br.
Fire Steels at 4 ℘ Br.
Files at 1 ℘ Br.
Wormes at 4 ℘ Br.
Nett Lines at 2 ℘ Br.
Awle Blades at 8 ℘ Br.
Twine at 1 Skeine ℘ Br.
Needles at 12 ℘ Br.
Fish Hooks at 20 ℘ Br.
Knifes at 8 ℘ Br.

Standard of Trade

Spoones at 4 ℘ Br.
Stockins at 1 pair ℘ Br.
Shooes at 1 pair ℘ Br.
Powder Horns at 2 ℘ Br.
Hatchetts at 2 ℘ Br.
Painted Boxes at 4 ℘ Br.
Shirts at 1 ℘ Br.
Combs at 2 ℘ Br.
Tobacco Boxes at 2 ℘ Br.
Ditto Tongs at 2 pair ℘ Br.
Pewter Buttons at 12 Doz. ℘ Br.
Rings at 6 ℘ Br.
Sispon at 2 ℘ Br.
Ice Chissells at 2 ℘ Br.
Scraper at 2 ℘ Br.
Leather Looking Glasses at 2 ℘ Br.
Tin Shoves at 2 ℘ Br.
Hawkes Bells at 16 ℘ Br.
Ostridge Feathers at 2 Plume ℘ Br.
Iron Wire Handcuffs at 1 pair ℘ Br.
Horns for mens Heads at 1 pair ℘ Br.
Yearne Gloves at 1 pair ℘ Br.
Turkey Reeds at 2 ℘ Br.
Carps Breeches at 3 Brs. ℘ pair
Broad Cloth at 12 Brs. ℘ Yard

Albany Fort America Anno 1706

Standard of Trade

Bayes at 1 Yard ℘ Br.
Flannel at 1 Yard ℘ Br.
Duffields at 1½ Br. ℘ Yards
Blanketts at 6 Brs. ℘ Blankett
Brandy at 4 Brs. ℘ Gallon
Arrow Heads at 12 ℘ Br.
Capps at 2 ℘ Br.
Mens Coats at 5 Brs. ℘ Coat Edged and 4 Brs. ℘ Coat Plain
Youths Coats at 3 Brs. ℘ Coat

Beaver being the chiefe Commodity we Receive in the Trade we therefore make it Standard whereby we Rate all other Furrs and Commoditys we Deale in for Trading

4 Martens as one Beaver
2 Otters as one Beaver unless they prove extrodinary Good & then we Rate them 1 ℘ Br.
2 Foxes as one Beaver unless they prove as above mention'd & then we Rate them 1 ℘ Br.
1 Catt as one Beaver
1 Moose Skin as two Brs.

Standard of Trade

1 Woolfe as one Br.
1 Black Bare as 2 Brs.
1 Cubb as one Br.
2 Quaquahatch as one Br.
10 ℔ Feathers as one Br.
1 ℔ Castorum as one Br.
2 Deare as one Br.
4 Fathom Netting as one Beaver
8 Pair Moose Hoofes as one Beaver

Figure 3.5. Albany Fort Standard of Trade, 1706. HBCA B.3/d/15 fos. 9d-10 (T16889).

Credit: Hudson's Bay Company Archives, Archives of Manitoba

commanders, had very little trust in them, and were always worried that they would steal from the company, either intentionally or by being careless.

While we did not have direct access to the headquarters' letters, we could imagine them as the post commanders referred to them in their replies. We also used the annotations provided by the editors of the correspondences' volumes as they contained a lot of information on the context, which was relevant to understanding the arguments. Moreover, the study of the East India Company correspondence by the communication scholar Kitty Locker provides supporting evidence. Similarly to the Hudson's Bay Company, the East India Company is a famous distributed company, which had its headquarters, composed of committees and directors in London, and posts all over India and Southeast Asia. It was created in 1600 and dissolved in 1874. Just as with the Hudson's Bay Company, the East India Company's letters were archived. The 48,000 volumes of the company survive, bound in large, red leather volumes in the India Office Libraries in London, Calcutta, Madras, and Bombay. In her analysis, Locker (1991) demonstrated how the letters between the management in London and the commanders at the different posts

TABLE 3.2 Expressing emotions in organizational letters

Emotional expression practices	Writing mechanisms	Effects on emotional expression
Criticizing and complaining	Objectifying	State and substantiate their criticism and complaints; the fact of writing them down has a cathartic function.
	Addressing and specifying	The nuanced presentation of their complaints and criticisms.
	Reflecting	Provide rationale for their criticisms and analysis of the situations they are complaining about.
Justifying decisions and actions: transforming negative emotions into positive ones	Objectifying	Reference to previous letters or accounts (i.e., objects or proofs).
	Addressing	Imagine the questions and criticisms that headquarters might have and providing them with answers; providing contextual knowledge to explain some of their choices and create a sense of a faithful employee.
	Specifying	Provide many details is crucial to explain actions, answer questions, and create a sense of trust.
	Reflecting	Provide rationale for their actions but also creating an image of a wise, trustworthy employee.
Seeking empathy and trust	Addressing	Convey to the management the sense of "real" people who suffer and want the good of the company; associate the headquarters with their emotions.
	Specifying	Provide detailed records and explanations to present themselves as trustworthy.

were, similarly to the Hudson's Bay Company' letters we studied, full of emotions. In particular, she showed how the committee and directors read every paragraph of every letter and were ready to criticize and ask for clarification for any points. In return, the commanders of the posts started being defensive and sometimes even aggressive.

Our analysis of the post commander's letters, replying to the headquarters, which we will discuss more in the following section, and the editors' notes reveal patterns similar to Locker's analysis: the Hudson's Bay Company headquarters complained of arithmetical mistakes, disobeying superiors' orders, not providing the merchandise expected, errors of judgment, and wasting of the company's money. The objectifying mechanism allowed writers to state and substantiate their criticism.

Specifying supports the detailed criticism of the headquarters. It is interesting to note that at one point[1] the headquarters started listing their points in their letters and asked the commanders to refer to each of these points. This became a tool for the headquarters to get replies to every single one of their requests and criticisms, thus creating a bureaucratic style in the correspondence.

For the headquarters, expressing their criticism was crucial because it symbolized their power; in fact, given the distance separating them from the posts, this might have been their only power. What was even more interesting to us as we were reading these letters, written in this strong, hierarchical context, was the criticism expressed, often in mild forms and through various practices, by the post commanders. Indeed, it could have been dangerous for the commanders because objectifying also implies fixing things, leaving a trace; yet because of the other mechanisms, such as addressing the reader, reflecting on the written word, and specifying details and nuances, they could ground their remarks and, in some ways, make them sound more mellow. Yet, they did criticize, although indirectly, by venting their frustrations with the headquarters—with their frequent criticisms, with their lack of understanding of the situation in the posts, and with the lack of resources.

In the following example, the commander criticizes as much as he ignores the orders of the headquarters (transmitted orally) in the absence of letters from the headquarters, which would represent the headquarters' opinion and authority:

> Whereas Mr Thoyts informs me that your honours have been pleased to appoint him to be an assistant to me, I find no such orders in your instructions, therefore shall continue with Mr Adams in that station till such time you are pleased to order otherwise. (Letter 31, p. 131)[2]

In this example, the recipient ignores the verbal "order" of taking Mr. Thoyts as an assistant. Because such an order has not been written down, it is not binding. The absence of the letter of appointment is a perfect (negative) example of the role of the letters as proof of an order, a decision, or accountability, which we will discuss in Chapter 4. In this case, the commander uses the letter as a means of expressing his disagreement with the appointment and of resisting the headquarters. In this case, the face-to-face conversation has no effects in the absence of a written order that would endorse it. On the contrary, it is the spoken word that is "dead" in terms of its effects; the only effect the oral words have is to generate a written letter that would ask for another written document that would have real effects. The commander is interpreting the written word as being more powerful than the spoken word to challenge and resist the order of the headquarters. Therefore, because of the distance and the infrequency of the letters, Mr. Adams would probably be assured to stay in place for one more year.

In some letters, such as the one below, post commanders comment on the lack of success of some strategies recommended by the headquarters, indicating that they always knew these actions would fail, suggesting that they, the commanders, were not asked or listened to:

> 19th and 20th. In answer to these paragraphs it was always our opinions little success can be expected from a trade with such savages, whose manners are so rude and barbarous as leave no hope of cultivating correspondence with. (Letter 66, p. 245)

In this excerpt, the commander uses the objectifying mechanism to refer to the criticism made by the headquarters ("in answer to these paragraphs") but also to refer to his opinion ("it was always our opinion"), which, had it been listened to, would have avoided the failure. He then justifies his criticism and makes it more acceptable by providing a long description (specifying). The expression "little success" reinforces the criticism indicated by "it was always our opinion." Here the writer distances himself from the reader and from the subject matter, effectively saying, "Don't blame me; this wasn't my idea."

In the following excerpt, the commander criticizes the decision of the headquarters of not providing him with enough resources for his men. After a long and detailed paragraph, with numbers to explain to the headquarters the context and to support his argument (specifying and addressing), he concludes:

> In short, Gentlemen, it is the only thing that your servants complain of and not a man of them but would abate 4 or 5 pounds of their wages, if they were sure of having a reasonable quantity of good flour, but the last year scares most of them, for I could not allow them more than 3 lb. of flour in the winter and added one lb. wheat to it when the long days came in. (Letter 1, p. 6)

The commander here uses a rhetorical strategy supported by the specifying mechanism to criticize the headquarters' decision of not providing enough flour. In another letter (which indicates that he has not been listened to), he threatens to increase his men's wages without the headquarters' approval. In the letter below, the criticism is even more explicit:

> Honoured Sirs, Your kind letter dated the 30 of May 1705, came safe to hand by Captain Grimington in the *Hudson's Bay* who thanks be to God arrived here in safety the 28th of August last, which was no small rejoicing to us after all the hardships we underwent occasioned by your not sending of ships the year before last with a supply; for which you have lost sufficiently by omitting the same [...] But this failure of shipping so continually, is the ruin of all, nay the very Indians now upbraid us with it, and tell us we have but a single ship once in two or three years which is really very hard. (Letter 2, p. 15)

A little later in the letter, Adam Blaye, the commander, refers to the letter (objectifying), which fortunately arrived but also provides context (addressing the reader). He even threatens to quit if the headquarters doesn't start changing this practice: "Gentlemen, if you send over ships yearly, a commander for the ship in the country, and men to defend the factory, I will continue in your services. But if not I will come home, God willing, with the next return of the ships" (Letter 2, p. 16). The commander supports his threat by reflecting—in the rest of his letter—on the implications of the lack of resources, which are cruelly needed, but also have a negative impact on the morale of the men, who all want to go back home. He also provides a rationale for his suggestion of sending two ships:

> I think it is much safer always to send two small ships rather than one large one by reason the voyage is very dangerous so that if one of them falls into any danger the other may assist them. I hope you will take care of that when you hear the great danger the *Hudson's Bay* was in the last voyage running aground a rock in the Straits when 'twas through providence they got off. (Letter 2, p. 16)

Given the organizational context of a strong hierarchy, we can surmise that statements such as the ones in the above quotes, especially the one where Blaye evoked the possibility of quitting, are fueled by strong emotions.

Writing's specifying and addressing mechanisms allow post commanders to be nuanced and careful in the presentation of their grievances. Writing down their complaints may have had an additional benefit for the post commanders. We know that when not addressed, negative emotions can have a strong detrimental influence on organizational atmosphere and on decision-making processes (Maitlis & Ozcelik, 2004). Hence, expressing negative emotions in organizational settings has, we argue, a cathartic function similar to the one discussed earlier in this chapter for individuals going through harsh or traumatic experiences (e.g., Lepore & Smyth, 2002; Smyth & Pennebaker, 1999). Writing in traumatic contexts is beneficial because it is asynchronous, allowing people to express things that they would not dare expressing face-to-face and to reflect on their experiences. The harshness of life in the posts—which we got glimpses of in some of the above excerpts—could be a traumatic experience, and the commanders might have felt better after expressing their complaints in writing.

Justifying Decisions and Actions: From Negative to Positive Emotions

Writing not only allows patients going through traumatic experiences to merely vent their emotions, it also provides them with the opportunity to self-reflect and articulate these emotions (Smyth & Pennebaker, 1999).

Similarly, our analysis shows that the commanders not only complained about their difficult situation, lack of resources, or lack of support from the headquarters, but they also reflected, explained, and articulated their positions to the management in London. In doing this, they also aimed to transform negative emotions into more positive ones through what we call justifying practices; that is, they tried to convince the headquarters that they did not do anything wrong, or at least did not intend to do so. Many of the letters of the commanders illustrate justifying practices. Justifying practices often rely on the objectifying mechanisms of writing, such as commanders referring to accounts, to the post journal, to previous letters, or to documents sent as proof with the letter, as in this excerpt: "And to let your honours see the manner that I have proceeded therein I have sent enclosed in the packet a copy of the instructions that I give from time to time to Mr. Napper and Mr. White to be observed by them in my absence at that building" (Letter 67, p. 258). Objectifying is used to refer to the commander's instructions to two of his subordinates as well as a proof for the way he manages the post. Absence of physical collocation is not the only reason for the profuse use of written instructions: Even among post managers, written material is used, most likely for reasons of traceability and clarity. In this letter, the commander attempts to justify his management style and prevent any criticism.

Objectifying is again central to the excerpt below, where the commander refers to a previous letter but also reflects on his action at the time, justifying the absence of written orders or contracts, called indents: "The reason why sent no particular indent the last time I wrote was that because I did not know but the goods that came over then" (Letter 1, p. 7).

In some cases, the commander not only refers to the facts and circumstances (which he explained in detail in the passage previous to the one quoted below), but also refers to his values and sense of ethics to ask headquarters to trust him (addressing and specifying):

> Thus may It please your honours has been my conduct in carrying on this affair, and if these circumstances is not sufficient to support me against the base report of indolence, a thing that I always had an aversion to, then must I fall, as to my grief I find myself under your honours' displeasure; for more than this I am not capable of doing, and yet I am very confident that no person could have done more to the dispatch and forwarding of this work than I have done. (Letter 67, p. 250)

In this emotion-laden passage, the commander is very open with his emotions, which he names directly (e.g., "aversion," "grief," and "displeasure").

Last, without replying to specific criticisms, the commanders of the posts enact the justifying practices by frequently referring to how obedient they are in following headquarters' orders. Addressing the managers at the headquarters is important in creating this sense of a faithful employee; the commander would sometime justify himself by imagining a possible criticism and the management's reaction: "I have sent an indent home of what is wanting, peradventure your honours may think it large" (Letter 9, p. 43).

The following letter illustrates a similar attempt, which aims to justify the actions prior to any criticism as well as to create trust:

> We sent letters to Churchill the first of this summer by Indians that went from hence last June, but have received no answers as yet: a copy of which letter you will find here enclosed, likewise a copy of letter received from Mr Myatt last fall 1727, with a copy of the answer sent this summer by Indians; and shall continue sending to your honours copies of all letters received, with the answers, whilst I am in your service, for I never had any clandestine dealings with any masters of your factories nor with your captains of ships nor with any tradesmen in England, for I dare them. (Letter 32, p. 34)

In this letter, the author tries to reassure people in London that they can trust him, that he "never had any clandestine dealings"; his main tactic to support his claim is to share with headquarters the copies of the letters he wrote to other posts. It is interesting to note that in this case, objectifying is intertwined with addressing: The writer, through the objectification process of writing, tries to create a context so that others imagine him as a faithful employee.

Because of the distance, the lack of contextual knowledge and the negative a priori that the headquarters had about the post commanders, it was very important for the latter to make sure that they transform the negative emotions of the headquarters—expressed in their criticisms—by justifying their actions. Some of the justifying also took place upfront by providing reasons for every action to prevent any questions or misinterpretations. Objectifying was crucial as it allowed the commanders to create a sense of objective proof for headquarters and often took the form of keeping copies of files and of letters sent to other posts. Specifying supported the justification strategy by allowing the commanders to provide many complex details about the post operations. Addressing was also important as it allowed the commanders to provide context to the managers in London but also to refer to what could be their feelings and interpretations when reading the letters.

Seeking Empathy to Build Trust

Apart from the accountability meant to prove their trustworthiness, post commanders also suggested empathy in order to build trust with the management. Two key mechanisms for suggesting empathy are addressing the reader and specifying. Hence, the commanders give headquarters a lot of details about the difficulties of everyday life:

> The place as we are come to is nothing but a confused heap of old rotten houses without form or strength, nay not sufficient to secure your goods for the weather, not fit for men to live in without begin exposed to the

> frigid winter. My own place I have to live in this winter is not half so good as our cowhouse was in the Bottom of the Bay. (Letter 8, p. 38).

The references to emotions also allow commanders to present themselves as "real people" who suffer but can also have positive emotions if they are rewarded:

> Honourable Sirs, by the arrival of Captain Christopher Middleton, commander of the *Hudson's Bay* frigate, in Albany road I received your honours' letters and herein do return you my humble and grateful acknowledgements for this so singular a favour, in my being appointed Chief Factor at Moose River. I now assure you it shall be my study faithfully and diligently in this so great a trust which your honours has been pleased to repose in me. (Letter 44, p. 171)

Another way in which post commanders tried to solicit empathy was by referring to their own emotions as they related to the managers' presumed emotions (addressing mechanism). The result is the creation of a shared set of emotions and expressing the identity of a "company man." For example, they empathize with the surprise or anger of the headquarters by noting how surprised and angry they themselves are:

> 2. What we observed to your honours concerning Mr. Bird in our last general letter, is all that we can answer on that head, and are sorry that there should be so many mistakes, but are equally surprised how they should happen.
> 3. We have according to our promise taken due care this year to avoid any mistakes. (Letter 79, p. 323)

The addressing mechanism plays a key role in supporting the commanders' attempts to associate with the headquarters' emotions. In another example, the commander replies to a criticism of the headquarters by explaining how afflicted he felt about the headquarters' unhappiness with his decision to pursue the rebuilding of a factory: "25th. It is a great affliction to us to find your honours so averse to our humble endeavours in carrying on the building" (Letter 66, p. 245). In later sentences, the writer reinforces the expression of affliction by highlighting how hard he and his men worked on this project, which, he notes, had been originally approved by the headquarters and started by his predecessor.

Trust is crucial to distributed organizing (Jarvenpaa & Leidner, 1999; O'Leary et al., 2002; Wiesenfeld, Raghuram, & Garud, 1999). It was particularly crucial in a company such as the Hudson's Bay Company, where there was only one letter per year exchanged and very little common knowledge of the context because the managers in London rarely had been to the posts in Canada, and many of the commanders had never been to the London office. In this context, it is not surprising to find, in headquarters' letters, an awareness that they had to trust the post commanders to act in the company's best interests. For example, the headquarters, in its 1682 instruction to one of the governors, wrote: "We must leave much to yours and the Captain's prudence and conduct, not doubting but [that] you will both do that which in your understanding [is] best for our service" (Rich, 1957, pp. 57–58, cited by O'Leary, Orlikowski, & Yates, 2002, p. 39). O'Leary et al. insisted on how the headquarters used the asynchronous communication to establish and express trust as well as control commanders' behavior.

At the same time, it was key for the post commanders to portray themselves as company men in order to keep their job. Most Hudson's Bay Company employees were Scottish, long servicing, and invested with great loyalty to the company (Hudson Beattie & Buss, 2003). Once they joined the company, they would spend most of the rest of their life working for the company and would never go back home. Our analysis shows how the only way for the commanders to build trust was by providing records and explanations of their actions, by justifying their decisions, and by showing by justifying their decisions, and by specifying their difficulties and even their suffering to show they cared for the company. This last element, the care of the commanders for their company, infuses many of the letters, and often takes the form of their portrayal of the difficult lifestyle, even the suffering they endured. In some ways, even through their complaints, they showed that they cared as they often referred to their interest in the company's success to ground their requests and complaints. Through these practices, the commanders affirm their identification with the company and participate in the construction of a sense of common identity, which is crucial in distributed organizations (Fayard & DeSanctis, 2010; Wiesenfeld et al., 1999).

Organizational research has recognized the emotional nature of organizations and organizational life (e.g., Ashkanasy, Zerbe, & Härtel, 2002; Elsbach, Sutton, & Principe, 1998) and our analysis of Hudson's Bay Company's organizational letters as well as Locker's (1991) analysis of the East India Company shows how letter writing was a key practice, enabling the expression of emotions—both negative and positive—and the development of trust in distributed organizations. If writing is a fundamental mode of communication for expressing emotions in distributed organizations such as the Hudson's Bay Company, its role in expressing emotions is even more crucial, and more visible, in personal correspondences, such as in the correspondence of Virginia Woolf, which we discuss in the following section.

EMOTIONAL CONNECTIONS IN PERSONAL LETTERS

If emotions are to be found in organizational letters, they prevail in personal letters; in particular, when the letter writer is a famous novelist such as Virginia Woolf.[3] In this section, we present our analysis of

Virginia Woolf's letters, which illustrate the extent to which writing can convey subtle and complex emotions.

The Correspondence of Virginia Woolf

Virginia Woolf (1882–1941) was a British novelist and essayist who is regarded as one of the most important modernist literary figures of the 20th century. Her most famous works include the novels *Mrs. Dalloway* (1925), *To the Lighthouse* (1927), and *Orlando* (1928), and the book-length essay *A Room of One's Own* (1929). Woolf was born into an upper-middle-class family in London; her father, Leslie Stephen, was a scholar and writer, and her mother was Julia Stephen. Woolf spent her first 22 years in the house she was born in, on 22 Hyde Park Gate, educated by her parents and music teachers. Neither she nor her sister went to school or university, but they grew up in an intellectual and free-thinking environment. Their father read literature to them every day; their mother had a salon, and many intellectuals of the time regularly visited the Stephen family. In 1904, Virginia and her sister, the painter Vanessa Bell, and their two Oxford-educated brothers moved to Bloomsbury, a neighborhood in Central London, and became the center of The Bloomsbury Group. This informal group of artists and writers, which included Lytton Strachey and Roger Fry, exerted a powerful influence over early 20th-century British culture.

Virginia Woolf is not only famous for her published novels and essays but also for her diaries and her correspondences. Apart from her diary, she wrote about five or six letters per day and hundreds of them over the years to her sister (Curtis, 2006). She started writing letters when she was six years old and wrote thousands of letters from 1888 to 1941, when she wrote her last letter to her husband, Leonard Woolf, before committing suicide (Sellers, 2000). Through her letters she developed and maintained multiple relationships with friends and loved ones, some living in England like her, others living abroad. Most of her correspondents kept her letters, which allowed their collection and printing after her death (Curtis, 2006).

Literary critics consider these letters as central to the understanding of Woolf's work and, in particular, her development of the stream of consciousness technique. Stream of consciousness is characterized by a flow of thoughts and images and corresponds to an individual's point of view. The text reads as an internal monologue, which evokes the continuous flow of the character's thoughts, feelings, and memories. In particular, the stream of consciousness that she first developed in her novel *Mrs. Dalloway,* to describe the typical hostess, had been explored and shaped in her letters (Trautmann Banks, 1990). Some critics even describe some of her letters as pieces of art in their own right (Sellers, 2000). The correspondence of Woolf explores many emotions related to loves, illnesses, achievements, and deaths on which she reflected and which she shared with her friends and family. Another noteworthy feature of these letters is the different tones she adopted for each individual (Curtis, 2006).

Figure 3.6. Portrait of Virginia Woolf, by Charles Beresford (1902).

Source: http://en.wikipedia.org/wiki/File:Virginia_Woolf_by_George_Charles_Beresford_%281902%29.jpg

Our analysis of Virginia Woolf's correspondence suggested three different practices supported by the mechanisms of writing that allowed her to express emotions and build relationships with her correspondents: articulating emotions, understanding emotions, and engaging in a true dialogue. Articulating emotions refers to the experience of the author who, as she writes down her emotions, reflects on them, and then organizes them. By objectifying her emotions, the writer deepens her understanding of her own feelings. Eventually, through objectifying and understanding their emotions, correspondents are able to engage in a true dialogue, as writing allows a personalized and individualized communication. Table 3.3 provides a summary of the writing practices and mechanisms as well as their effects on emotional expression that we found in our analysis of Virginia Woolf's correspondence.

Articulating Emotions

Many of Virginia Woolf's letters provide examples of the stream of consciousness technique that allowed her to articulate her emotions by

TABLE 3.3 Expressing emotions in personal letters

Emotional expression practices	Writing mechanisms[1]	Effects on emotional expression
Articulating emotions	Objectifying	Allows Virginia Woolf to articulate in a vague, open form, feelings and thoughts; by writing them down, gives them an existence and provides a cathartic function.
	Addressing	Writing her thoughts in the succession in which they came to mind, without forcing a structure, as if she were talking aloud to her friend.
	Specifying	In some cases, writing forces the writer to be specific and articulate and name often-confused emotions.
Understanding emotions	Objectifying	Allows multiple readings, which can sometimes increase emotions even when they are painful from the beginning.
	Addressing	Forces the writer to organize confused emotions in order to express them (knowing that some are not expressible) and to also take into account the reader's expectations to make oneself comprehensible.
	Reflecting	Gives one the opportunity to relive the emotions associated with the relationship, go back to old letters to make sure they were understood correctly, and to nuance that understanding.
Engaging in a true dialogue	Objectifying	Letters can be kept and read many times; they can become a symbol of the dialogue and of the relationship, to which one can return as to a soothing presence.
	Addressing	Personalized and individualized communication; plays an important role in the establishment of a meaningful exchange between correspondents. Both the writer and the recipient are situated in a moment where both can imagine the other.

[1]Here we only discuss the mechanisms directly at play in a particular practice.

staying close to them—by mimicking them, in a way. The following is an excerpt of Virginia Woolf's letter to her friend the musician Ethel Smith. Ethel had received a negative critique by the music critic of the *New Statesman* magazine, W. J. Turner, to which she replied three times in print, each time reacting more strongly. Virginia Woolf, in a first letter, had shared with Ethel her thoughts about Ethel's strong replies. Ethel was hurt by what she thought was a critique by Woolf. Virginia Woolf replied, trying to clarify her position:

> And I respected him; and I respect you. Only I think you don't altogether realize how, to the casual onlooker you seem exaggerated—how it strikes an outsider. I think sometimes you let the poison ferment. Never mind. I can't altogether lay hands on my meaning. The other thing that interest me much more [...] Oh, and when I write to you, I put it off till the end of the day. (May 12, 1931)[4]

This excerpt illustrates the reflecting mechanism "in action," as when Virginia Woolf tries to grasp her impression of Ethel: "Never mind. I can't altogether lay hands on my meaning." She also shares her reflection with her reader: "Only I think you don't altogether realize how, to the casual onlooker you seem exaggerated." By sharing her reflections with Ethel, Virginia Woolf is engaging in a dialogue with her reader. The dialogic nature of the stream of consciousness is also illustrated by her switching among various topics, which allows her to share with her correspondent her different perceptions and feelings.

This example shows how Woolf follows the flow of her emotions, writing what she feels, in the succession in which these come to mind, without forcing a structure, as if she was talking aloud to her friend Ethel. The addressing process therefore is at the core of stream of consciousness writing. Writing has fixed on the page these fleeting, vague, not fully formulated feelings and thoughts, and this provides a window into the writer's state of mind, not only through the words used but also through the style itself.

If writing allows one to articulate in a vague, open, form one's feelings and thoughts, it also, in some cases, forces the writer to be specific and articulate and to name often confused emotions. For example, Woolf writes to her nephew that, because she was sick, she was unable to write to him and to therefore express her emotions:

> My brain was packed with close folded ideas like the backs of flamingoes when they fly south at sunset. They are now all gone—a few grey draggled geese remain, their wing feathers trailed and mud stained, and their poor old voices scrannel sharp and grating—that's the effect of typing. (February 17, 1930)

Here, through the use of metaphors and the writing process, Woolf articulates her emotions, giving them existence and sharing them with her correspondent. Furthermore, writing is cathartic, as it helps her liberate herself from the disturbing ideas that were cluttering her mind before she had starting expressing her feelings in writing: "They are now all gone [...] that's the effect of typing."

Letters provide a way to express the self because they give the writer flexibility to follow the flow of his or her emotions, assess a feeling, and also to express contradictions and move from one topic to another (Rousset, 1966). In a sense, writing letters is like talking aloud, objectifying one's emotions without necessarily organizing them but following one's stream of consciousness. Yet, writing emotions as one experiences them is only one aspect of letter writing. In the writing process the writer can also become aware of additional emotions and nuances. Thus, complex emotions can be expressed through writing.

Understanding Emotions

Writing in general, and letters more specifically, can be seen as the antithesis of an oral dialogue because of the asynchrony and lack of immediate feedback it involves. However, it seems that in some cases, this asynchrony allows the correspondents to reveal more about themselves. Indeed, both the writer, who can draft a letter several times, and the recipient who can read it multiple times before replying, have the opportunity to relive the emotions associated with the relationship, to go back to old letters to make sure they understood them right, and to give nuance to that understanding.

This opportunity to deepen understanding is beautifully described by Virginia Woolf in a letter to Gerald Brenan, a young writer and friend. On October 4 Woolf writes that she has just reread a letter that he sent her on July 11. Then she writes:

> And if it should occur to you one night to attempt this curious effort at communicating what Gerald Brenan thinks on his mountain top, then I will first read it through very quickly, at breakfast, and come upon it a little later, and read it again and try to amplify your hieroglyphs. (October 4, 1929)

The *objectified nature* of the letter allows for multiple readings, which can sometimes intensify emotions even when they were painful from the beginning. The absence of the other person and of immediate feedback can help manage the emotions and allow reflection before replying.

Letters are particularly powerful because they are written to another person; thus, they allow the writers to develop their selves through the relationship with that other person. While writing a letter, we articulate our inner self for our friend—organizing on paper our confused emotions and trying to name feelings—and by so doing we clarify our emotions for ourselves and thus develop an understanding of them. Therefore, the addressing mechanism facilitates the process of expressing as well as understanding emotions. Hence, in the following letter to her friend Ethel Smith, Virginia Woolf explores her feelings in order to clarify her understanding, as well as Ethel's:

> No Ethel, dear, no; I didn't make my meaning plain. I wasn't alluding to any particular instance of misunderstanding [...] this instance—your behavior

> about critics and your music—doesn't seem to me of importance. That is, if I imagine, by imagining you as a whole,—with all your outriders and trembling thickets of personality, exactly why you do it; and sympathise, and admire; and feel the oddest mixture of admiration and pity and championship. (May 12, 1931)

This excerpt illustrates how addressing Ethel—"if I imagine you by imagining you as a whole"—helps Virginia reflect—"I didn't make my meaning plain. I wasn't alluding to any particular instance of misunderstanding"—articulate, and deepen her understanding of her emotions about her friend.

Expressing one's emotions is not easy because feelings are often confused and subtle. In fact, a common statement when trying to do so is "words cannot describe how I feel," and even expert writers such as Woolf express this frustration in some of their letters. This seems to be supported by studies claiming that emotions tend to be perceived and expressed nonverbally rather than verbally (Ekman, Friesen, & Ancoli, 1980). This may be the case. At the same time, this argument does not imply that writing cannot support the expression of strong and nuanced emotions. Thus, we argue that when one does try to find the words, as suggested by research on expressive emotions (Lepore & Smyth, 2002; Pennebaker, 1990; Smyth & Pennebaker, 1999), a lot is achieved, both in terms of the writer's own understanding, as we showed previously, and in terms of the reader feeling connected to the writer. While writing is often seen as making the expression of emotions harder because of the absence of the other person, the inability to use nonverbal behaviors, and the lack of shared context, our analysis suggests that writing can help in expressing emotions because one needs to organize confused emotions in order to express them (knowing, however, that some are not expressible) and to also take into account the reader's expectations to make oneself comprehensible. The absence of the other person and of a shared context forces the writer to address the recipient to express and explain his or her feelings. Thus, writing mechanisms, in particular objectifying and addressing, support the expression of emotions and therefore, albeit indirectly, the building of relationships.

Engaging in a True Dialogue

Writing is often seen as a universalizing medium (Auroux et al., 1996; Plato, 1940a; Rousseau, 1763/1959), which does not allow the intentionality required for the development of relationships. However, as noted in Chapter 2, the addressing dimension of writing makes it intentional (Altman, 1982; Derrida, 1987; Ong, 1982/2002). In fact, letters are always intentional and addressed to someone: They are written to someone, and the writer not only imagines the person he or she is writing to and the recipient's reactions but also provides the recipient with some contextual details about the experience and state of mind of the writer. Letters allow correspondents to engage in a true dialogue[5] where they can exchange emotions and develop their relationship. This dialogue is exemplified by famous cases of relationships, which developed through letters and sometimes mostly existed through letters, such as Elizabeth Barrett and Robert Browning or Franz Kafka and Milena Jesenská. While Virginia Woolf was not involved in a similar relationship, she did have continued letter exchanges with the most important people in her life—such as her sister, Vanessa, her friend Vita, and her husband, Leonard. These relationships developed and evolved not only through face-to-face encounters but also through written exchanges, which offered Virginia a space to express her complex emotions, particularly during periods where she was sick and could not see people.

The addressing mechanism, which links the reader and the writer, plays an important role in the establishment of a meaningful exchange between correspondents. This writing mechanism is wonderfully described by Virginia Woolf in a letter to Gerald Brenan:

> Suppose you are in the mood, when this letter comes, and read it in precisely the right light [...] then by some accident there may be roused in you some understanding of what I, sitting over my log fire [...] am, or feel, or think. (October 4, 1929)

This excerpt shows how letters situate both the writer and the recipient in a moment where both can imagine the other. Addressing the reader is a central process that is crucial in the development of a relationship because it allows the correspondents to engage in a real dialogue through which they can build and strengthen their relationship.

Writing, and letters, more specifically, can be seen as the antithesis of an oral dialogue because of asynchrony and lack of immediate feedback. However, it seems that in some cases, this asynchrony allows the correspondents to reveal more about themselves. Trautmann Banks stressed the importance of the (imagined) reader in Virginia Woolf's letters and claimed that we know more about Woolf (and thus that she expresses more) through her letters than through her diaries:

> Thinking alone as opposed to writing, may remove us even further from our personalities in that there is less of a context. [...] When we write a spontaneous personal letter, we *inhabit* our selfhood and use it to reach out to another person. It is in relationship that we form our identities as babies. It is in relationship that we continue to form them. (Trautmann Banks, 1990, pp. xiii–xiv)

The fact that letters reveal more of the personality of their writers than diaries do has been noted in other cases, such as the correspondences and the diaries of Jean-Paul Sartre and Simone de Beauvoir (Poisson, 2002). Indeed, as suggested previously, letter writing is a conversation, a dialogue between the writer and his or her imagined reader, just like we saw with Virginia Woolf and Gerald Brenan. In this dialogue, the writer is, in a sense, less self-conscious, and some elements of her personality might be revealed in her "interactions" with the imagined recipient of the letter. Thus, we surmise that writing allows a personalized and individualized communication in which correspondents are able to develop a true dialogue and strong relationships.

Moreover, as letters can be kept without being read immediately or can be kept and read many times, as many times as wanted, they also become a sign of the dialogue and of the relationship to which one can return, over and over, as a soothing presence. Woolf wrote to her close friend Violet Dickinson:

> My Beloved woman, your letters come like a balm on the heart. I really think I must do what I never have done—try to keep them. I've never kept a single letter all my life—but this romantic friendship ought to be preserved. (May 4, 1903)

In this way, letters become a proof of the relationship, even in the absence of the loved person. Incidentally, Woolf did keep her letters. As she wrote, letters become a trace of the exchange (through the articulation process they can be kept and reread) and therefore become a sign of the relationship.

Our analysis of Virginia Woolf's correspondence indicates that letters can express strong and nuanced emotions. Of course, letters are not the only genre that can convey powerful and subtle emotions. All of us who have wept while reading *Beloved, David Copperfield, Wuthering Heights,* or the more recent *The Life of Pi* or *Disquiet* can attest to the power of writing to move hearts. Interestingly, a particular genre of novels, epistolary novels, in which the protagonists exchange letters, are particularly moving at they allow readers a direct glimpse into the protagonists' tribulations. The contemporary novel *Chinese Letters* by Ying Chen is a perfect illustration, as it shows us through 57 letters exchanged between two lovers, Yuan and Sassa, and two friends, Sassa and Da Li, the difficulty of leaving, of long distance relationships, and the evolution of feelings (see the sidebar Emotional Expression in Epistolary Novels).

Letters also facilitate the meeting of two subjectivities and the development of strong relationships. One might argue that a limitation of our argument is Virginia Woolf's talent as a writer. True, her letters are incredible examples of craftsmanship in the expression of emotions. Nonetheless, the power of writing to express emotions and build relationships is also enacted through letters of nonfamous people. For example, letters sent by families to the sailors, laborers, and tradesmen who were employees of distributed organizations such as the Hudson's Bay Company are examples of personal letters, in which, many times, emotions were expressed in a most poignant way (Hudson Beattie & Buss, 2003). Similarly, the letters of young French soldiers in the trenches of the First World War told their fiancées, families, or war godmothers about their experience—their fear, their pain, and their happiness (Darrow, 2000).

EMOTIONAL EXPRESSION IN EPISTOLARY NOVELS

While letters have been the means for people to communicate since the advent of writing, in the 17th century, letters make their grand entrance on the literature scene, with the invention of the epistolary novel (Rousset, 1966), which is a novel written as a series of letters. Thus, novelists, who are experts in expressing nuanced and complex feelings, chose letters as the best vehicle to express emotions. Hence, *Letters of a Portuguese Nun* (1669/2007), which is considered to be the first epistolary novel, is an unequaled model of the expression of passionate love (Rousset, 1966). In fact, for a long time, these letters were the model of the ideal love letter (Rousset, 1966) and show how the epistolary style fits the intensity of passionate love and allows for the immediate and spontaneous expression of changing emotions. For example, one letter states: "I don't know neither who I am, nor what I'm doing, nor what I desire. I'm torn apart by a thousand of opposite movements" *(Letters of a Portuguese Nun,* 1669/2007, 3rd Letter). This excerpt indicates some of the fluctuations and contradictions of love, which are expressed in the whole letter, through contradictory sentences, in which wishes expressed in one sentence are taken away in the next. In fact, these letters were so expressive that for a long time people believed that the letters were really written by a nun and were not a work of fiction.

The main reason for the success of epistolary novels resides in the fact that the reader can sense more of the passion when it is rendered by the hero or heroine rather than by a narrator (Montesquieu, 1754/1973). Letters allow us to express things in our own name and in the present, thus bringing the reader closer to the author's feelings. The reader becomes a contemporary to the action. He or she lives it as it happens. In other words, letters allow a direct contact with the situation and feelings of the author. In a way, reading an epistolary novel is not unlike watching a play, in which we see characters evolving as they are moved by their passions. The reader of the epistolary novel can thus attend—without the intermediary of the novelist's voice—to the dialogue between the letters' writer and reader.

Therefore, in epistolary novels and in letters in general, we get to see the expression of emotions and their variations in all their subtlety and force. Through the addressing process which is at the heart of epistolarity, letters facilitate a dialogue between the writer and the reader and support the development and the deepening of their relationship.

What our analysis shows is that writing is not an impoverished medium when it comes to emotions. Although writing is seen as limited in its ability to express emotions and develop relationships because it objectifies fleeting states of mind, the objectifying mechanism is in fact crucial for emotional expression and relationship building. The specifying mechanism of writing allows the writer to both mimic the flow of emotions—reproducing a stream of consciousness—and articulate subtle and complex emotions. Once written down, emotions can be reflected upon. The asynchronous nature of writing, which is also seen as a flaw, can in fact give the writers the freedom to reflect and think about the wording. It also alleviates the emotional work sometimes involved in face-to-face situations, especially when it comes to expressing negative emotions. Last, although writing is often seen as universal and deindividuating, it is in fact always personal, and any writing, especially letters, is always addressed *to someone.*

WRITING AND EMOTIONAL EXPRESSION AND EXPLORATION

While writing is often criticized because of its fixedness and its impersonal nature, our analyses of letters between the post commanders of the Hudson's Bay Company in Canada and their managers in London, and Woolf's personal letters, show how writing, through objectifying, addressing, specifying, and reflecting, enable the expression of emotions (negative and positive) and the development of interpersonal bonds in organizational and personal contexts. Writing allows us to turn our experience into words and provides us with a space to articulate and express our emotions. Virginia Woolf's stream of consciousness is a beautiful example of this phenomenon. It illustrates how writing is a creative process, which does not freeze emotions but supports their articulation. It is also through writing that employees of the Hudson's Bay Company were able to express their frustrations and voice their criticisms and suggestions while still maintaining a sense of hierarchy and respect of order.

Building relationships usually implies having face-to-face interactions, as people tend to believe that the reciprocal trust needed for the development of strong interpersonal bonds is not possible in the absence of social and nonverbal cues. Yet, organizational letters such as the Hudson's Bay Company letters or personal correspondences such as Virginia Woolf's correspondence illustrate how people can develop trust and relationships through writing.

CHAPTER

4

Knowledge Development Through Writing

> Probably [...] the whole scientific enterprise with the three of us would never have worked out if we were all sitting in one place. Cliff found this way of working, with me located miles away, to be just about the right level of controlled interaction for him to flower.
>
> —Alan Newell, about his distance collaboration with Herbert Simon and Cliff Shaw on developing artificial intelligence, in *McCorduck* (1979, p. 144)

Writing and knowledge are intrinsically connected in Western societies[1]; you just need to think of the wealth of knowledge preserved in scrolls, books, articles, and libraries. The Royal Library of Alexandria in Egypt has become a symbol of knowledge itself. This is not surprising. For example, during antiquity, it functioned as a major center surprising: since its construction in the third century BC until the Roman conquest of Egypt in 48 BC it functioned as a major center of scholarship. It was charged with collecting all of the world's knowledge. It is estimated to have stored at its peak 400,000 to 700,000 parchment scrolls. While there is no certainty on when and how it was destroyed, its destruction has been described as a terrible loss, many original texts having disappeared with the library.[2]

The imagery of libraries, books, and other forms of written documents as being closely associated with knowledge goes hand in hand with the view of writing as a recording tool, which helps create an "artificial memory" (Goody, 1987) and develop a large repository of discoveries, theories, and inventions across centuries and geographies. Writing as a recording tool has also been central in building an organizational memory, a "repository of organizational information" (Walsh & Ungson, 1991, p. 81). While distributed between brains and paper (Pondy & Mitroff, 1979), organizational memory rests heavily on the keeping of records and files (Yates, 1989), especially in distributed organizations.

While it is usually accepted that writing as a recording tool has supported knowledge sharing in distributed communities and organizations, we argue that writing is not only useful in recording and preserving information, but it is also immensely creative and has played an important role in knowledge development (Searle, 2010; Wolf, 2008). The correspondence of the French philosopher, mathematician, and scientist René Descartes (1596–1650)—who is sometimes referred to as the "Founder of Modern Philosophy"[3] and the "Father of Modern Mathematics"—with Princess Elisabeth of Bohemia,[4] which lasted from May 16, 1643, to December 4, 1649, is a wonderful example of how creative writing can be. Princess Elisabeth became acquainted with Descartes in 1639. Descartes, at her request, was made her teacher in philosophy and morals, and in 1644 he dedicated his *Principia* to her (Descartes, 1953c). Elisabeth's philosophical correspondence with René Descartes lasted for seven years, until his death in 1650. In this correspondence, Elisabeth asked numerous questions about Descartes's ideas in mathematics and metaphysics, pushing him to elaborate and develop his thinking. In particular, Elisabeth asked Descartes to explain what is the third "primitive notion," the body–mind composite (the first two primitive notions being the mind and the extended substance, or body) that he defined at the end of the *Méditations* (Descartes, 1953a). Her questions led him to acknowledge the impracticalities of his definition of the interactions between the body and the mind and forced him to rethink the body–mind relationships and their consequences when it comes to moral issues. Through this process, Descartes developed his thinking on morals. Therefore, when Queen Christina of Sweden asked Descartes to send her a treaty on his moral philosophy, he transmitted her copies of part of his correspondence with Princess Elisabeth, indicating that they were the draft of his moral philosophy. This draft became the *Passions of the Soul,* published in 1649, one of Descartes's main works on moral philosophy, which he dedicated to Princess Elisabeth (Beyssade & Beyssade, 1989).

Not only does writing enable the expression of complex thoughts or equations that could not be achieved orally, but it also helps create new entities such as laws and organizations, in particular in distributed contexts. For example, the Hudson's Bay Company, the trading company we introduced in Chapter 3, relied extensively on writing to function as a distributed organization as it was the only way for its employees to share their knowledge and to coordinate. In this chapter, we argue that writing's creative nature plays a central role in knowledge sharing and development. In particular, we show how the four mechanisms of writing—objectifying, addressing, reflecting, and specifying, which are central to the creative power of writing—support the development of two types of knowledge: operational knowledge in the form of organizational

TABLE 4.1. Writing mechanisms and their role in knowledge sharing and development

Writing mechanisms	Practices	Effects on knowledge sharing and development
Objectifying	Referring to a previous letter or document (that the writer have written or have received), e.g. mentioning that the letter has been read, shared or discussed. Referring explicitly to the act of writing down, e.g. "it took me time to reply in order to present my ideas clearly ... " or "as I'm finishing writing" Presenting hypotheses or summary of different events that took place.	Allows writers to share and refer to knowledge. Facilitates discussions and reflections.
Addressing the reader	Providing contextual information, e.g. on when the letter was written or received, or elements in the life of the writers that might explain better the decisions they took. Referring explicitly to the reaction of the reader, e.g. "you might disagree…" Providing extra information to clarify an idea that might be misunderstood, e.g. "in case you thought, here is what really happened or what I really meant".	Facilitates understanding as the recipient is provided with contextual information and the writer articulates the argument while aware of the potential questions the reader might ask. Allows knowledge to be more specific, and helps in the interpretation of knowledge.
Reflecting	Referring to the time taken to write or read a letter and to the interpretative process, e.g. "I received your letter a few days ago but I wanted to think before replying…", or "If I understand your question well…" Referring to the reading of the letter and questioning it, e.g. "as I was reading your letter, I came to wonder…" Referring to multiple readings of the letter.	Facilitates and enhances understanding. Allows the development and evolution of more complex knowledge.

(Continued)

TABLE 4.1. *(Continued)*

Writing mechanisms	Practices	Effects on knowledge sharing and development
Specifying	Developing long lists of events, objects, or people. Using complex sentences. Providing a lot of details and / or nuances.	Allows writers to articulate knowledge. Allows writers to develop and share subtle and complex knowledge.

memory in the case of the Hudson's Bay Company and scientific knowledge in the form of a new theory by Albert Einstein and Élie Cartan. Both of these cases illustrate how writing, thanks to its four mechanisms, enables the creation and sharing of complex knowledge (Table 4.1 shows the writing mechanisms and their effects on knowledge sharing and creation).

PERSPECTIVES ON THE ROLE OF WRITING IN KNOWLEDGE SHARING AND DEVELOPMENT

Historical examples of distributed organizations such as the Catholic Church (King & Frost, 2002), the Hudson's Bay Company (O'Leary, Orlikowski, & Yates, 2002), and the East India Company (Locker, 1991) or distributed scientific communities such as the Republic of Letters (Collins, 1998), which we will discuss in Chapter 5, highlight the importance of writing for knowledge development. The scientific communities of the 17th and 18th centuries, for example, could not have existed in the absence of the intense letter exchanges among intellectuals and scientists who lived all over Europe (Collins, 1998; Crane, 1972). Members of these "invisible colleges" (Crane, 1972) rarely met in person and were able to exchange ideas, develop and refute theories, and build personal relationships thanks to written communication (Bazerman, 1988; Collins, 1998; Crane, 1972). As discussed in Chapter 2, writing has had a particular influence on the development of scientific knowledge and is intrinsically related to knowledge creation and its sharing. Indeed, "the move in the direction of 'science' had much to do with the uses of writing itself" (Goody, 1987, p. 75) because writing allowed the recording of observations and then their precise comparison, which enabled generalizations, the testing of hypotheses, and, more generally, the formalization of knowledge (Goody, 1987).

Writing continues to hold a central role in contemporary organizations for which knowledge creation and sharing are crucial (Kogut & Zander, 1992; Leonard-Barton, 1995; Powell & Snellman, 2004). The growth of virtual organizations, communities, and teams has been accompanied by the need to express ideas and to coordinate via written communication. In this context, the reliance on writing is such that many groups and communities communicate extensively, and sometimes almost exclusively, through writing. Thus, members of scientific laboratories located in different countries use primarily writing (in the form of e-mail with attached documents that are shared and sometimes collaboratively written) to collaborate with faraway colleagues (Schunn, Crowley, & Okada, 2002; Walsh & Bayma, 1996). Also, the free/open source software communities develop complex software systems while communicating almost exclusively through writing (Lakhani & von Hippel, 2003; Moon & Sproull, 2002). Recent studies show the increasing reliance on e-mail to communicate in organizations with people at different locations but also with people sitting only a few meters away (Byron, 2008). Thus, we are writing more and more, even as the explosion in communication technology keeps adding new means of communication to our tools.

In spite of the increased reliance on writing for knowledge sharing and development, scholars who have studied knowledge sharing and development have not focused on the role played by writing in this process. The literature on knowledge transfer grew from an interest in the operations of large distributed multinationals and has focused on how to motivate the transfer of knowledge, which theorists assume has been created and coded in a way that can be transmitted (Kogut & Zander, 1992; Zander & Kogut, 1995). The theoretical and empirical importance of codification in these studies suggests that knowledge transfer largely rests on the use of written knowledge. However, the role of writing itself, as a fundamental communication mode, and its contribution in the transfer of knowledge has not been examined.

Other scholars interested in the sharing and development of knowledge have taken a more process-oriented approach and have highlighted that knowledge is an emergent and continuous accomplishment (Giddens, 1984; Orlikowski, 2002; Schön, 1983). These studies insist on the fact that knowledge does not exist as a fixed entity, but that it has to be revived each time it is used by specific people in order to be adapted to their needs and situation (Hutchins, 1991; Lave & Wenger, 1991; Suchman, 2007). For example, Lave and Wenger built upon studies of Yucatec midwives and Vai and Gola tailors to argue that they learned how to become midwives or tailors by becoming part of the community of experts, by observing them and becoming part of the group as individuals and professionals. These studies insist on the emergent understanding taking place in the interactions and in the participation of the community.

The importance of writing in the development of knowledge is acknowledged by some scholars in this perspective, but it is not clearly stated. Take, for example, the studies by Orlikowski and Yates (Orlikowski & Yates, 1994; Yates & Orlikowski, 1992) of the emergence and development of various genres of written communication in a collaborative project of software development. The focus on various genres of written communication as being central to the collaborative development of new software is an acknowledgement of the role of writing in knowledge development. Still, these studies are mode agnostic and thus stop short of examining how a specific mode, oral or written, supports the creation, development, and sharing of knowledge.

More profoundly, both approaches—the knowledge transfer literature and the process perspective—consider that tacit knowledge (i.e., knowledge that is not articulated and that, in fact, cannot be articulated)—are key to developing knowledge. Because they see writing and tacit knowledge as opposite, they therefore (implicitly) assume writing is appropriate only for recording explicit knowledge such as data and conclusions. In contrast, writing is seen as lacking when the knowledge to share is complex and/or ambiguous and is even more inadequate when it comes to creating rather than just recording knowledge.

ORGANIZING AT DISTANCE: SHARING KNOWLEDGE WITHIN THE HUDSON'S BAY COMPANY

The Development of Organizational Memory

A key challenge for distributed organizations is to develop and maintain an organizational memory that can be used as a reference by different members of the organization. This organizational memory is not a fixed entity but a shared knowledge base, including inventories, records of past actions, and decisions and rationales for these actions and decisions, which are constantly reinterpreted by the different members of the organization as they are doing their work. Our analysis of the Hudson's Bay Company letters showed how writing supported three main practices that are essential to the enactment of an organizational memory: creating a repository to archive not only the different inventories but also the different buying and selling strategies; building accountability to develop trust with the management in the headquarters; and providing context to help the headquarters understand the work conditions and possibly the rationale behind some decisions. A trading organization like the Hudson's Bay Company might be perceived as using writing mostly for recording and sharing knowledge. By writing down their inventories and explaining their buying strategies, commanders of Hudson's Bay Company were obviously recording their actions; however, through sharing their practices and problems with other commanders and with the managers, they also created a set of best practices that can be shared with other posts and used by the headquarters to develop procedures. As we

TABLE 4.2 Development of an organizational memory through writing

Knowledge practices	Writing's mechanisms	Effects on organizational memory development
Creating a repository	Objectifying	Share practices and explain rationale for actions
	Specifying	Provide detailed accounts of employees and activities; build a common ground.
Building accountability	Objectifying	Writing down creates a proof or reference.
	Reflecting	Review actions and reasons for actions as a way to show one's trustworthiness.
	Specifying	Explain actions in great details to show that one can be trusted.
Providing context	Addressing	Provide background information on the context of life to help the management in London imagine the situation and understand the decision of the post commanders.
	Specifying	Provide many details on the context and justify local decisions.

will highlight below, all four mechanisms of writing—of course, to varying degrees and in particular combinations—support the development of an organizational memory. (Table 4.2 summarizes the three practices and the role of the main writing mechanisms for each.)

Creating a Repository

The Hudson's Bay Company kept detailed records of all its activities. The main source of information for the headquarters in London and the different posts were the general letters addressed to the company's London committee from each post and a fort journal. The letter was generally 4 to 6 pages long (but some were up to 10 pages) and often had an inventory attached to it. It contained answers to the instructions or questions that the ships brought from London yearly and each post's story of the problems it was facing. There were also letters exchanged among the posts. Through these letter exchanges, post commanders learned about their coworkers' (whom they saw rarely, if ever) practices and sometimes adopted them; they also reflected on their own practices and, when needed, evolved them. In several letters, commanders referred to letters they had received from another commander with information on their situation and compared it with their own situation (e.g., Letter 9, pp. 40–46; Letter 60, pp. 222–225).[5] In another example, a commander explained why he was not going to depart from some of his predecessors' practices. He wrote that he would "not be wanting in maintaining the same correspondence with the leading Indians as my predecessors have done as well for the increase of the trade as to get of them information from time to time of the French's proceedings" (Letter 2, p. 17). The commander effectively stated that maintaining a correspondence with the local chiefs was a valuable practice that he intended to continue and expand.

Over time, references to other posts' letters became more frequent. We also noticed an evolution in the style of the letters, which became more structured, thus indicating a strategy to formalize the information shared in these letters. The early letters are long letters with references to the points made by the headquarters (e.g., "As for the repairing of the *Beaver* sloop this fall and next spring…") but starting from Letter 61, written in July 1737, commanders started numbering their points explicitly (e.g., "3. On the 4th of April last I sent twenty-eight labouring hands with bag and baggage," Letter 61, p. 227). One commander suggested that it was a rule imposed by the headquarters: "And according to your honours' directions in the 49th paragraph I shall take care to answer your honours' letters in a methodical manner paragraph by paragraph as they shall be numbered" (Letter 67, p. 255).

The annual letters are full of details regarding Hudson's Bay Company employees' actions: invoices, objects shipped to or from London and their values, and tables listing the names and wages of newcomer and of employees returning to England. These long lists and inventories demonstrate the way the objectifying mechanism (i.e., the act of expressing, writing down an idea or an emotion that can be then reread, reflected upon, modified, and shared with others) allows this information to be shared with the headquarters. Moreover, it also shows how the specifying mechanism (i.e., being precise and giving details as the writer tries to articulate clearly his or her points and answer the reader's possible questions) supports the development of detailed accounts of activities. It would be hard, if not impossible, to imagine the effective functioning of a distributed trading organization in the absence of these shared letters.

Yet, these letters provide not only long inventories, but they also explain how and why things are done, or cannot be done. As the annual letter was nearly the only means of communication with the headquarters, the commanders took pains to explain the decisions they made during the year, as illustrated in the following excerpt about gun trade:

> And I found the Governor when I came here for the same thing, and I told him my orders was to send them home.…I told them I could not nor would not take any charge of them, which at last we did agree and came to a price…I bought all they had so you will be no losers by it.…All the guns as is bought, I have computed the charge with the locks and stocks and other mounter, will not stand me in above twelve shillings each, reckoning the armourer's wages to fix them. (Letter 6, p. 35)

This excerpt shows how writing establishes a common ground for the commander and the headquarters by providing detailed information on events, prices, and merchandise as well as on the commander's decision process. In current distributed organizations, some of this information might be attached as a spreadsheet instead of the long, handwritten list added to the letter or inserted in the body of the letter, but the purpose is the same. This example shows how specifying is key for knowledge sharing in a company where the headquarters wanted to know all the details of the post operations. At the same time, by providing such a detailed account to the headquarters, the commander signals that he can be trusted. Thus, letters become more than just a database useful for accounting purposes—they become a rich resource of practices developed by the commander. They therefore play a key role in supporting organizing at distance as they provide headquarters with crucial knowledge to coordinate among distant sites, keep track of activities, and thus help decide what to do and how to do it. They also provide the headquarters with detailed accounts of each commander's activities and might be used to determine rewards based on results.

Thus, mostly through objectification and specifying, letters exchanged in distributed organizations support the construction of an organizational memory understood as stored knowledge (Moorman & Miner, 1998) including not only inventories but also explanations of practices and actions taken by different actors.

Building Accountability

Because writing provides a trace of past actions, letters can be used as proof of one's trustworthiness in distributed organizations whose employees often have never met each other and where trust might not be very high. In particular, smuggling, a private trade forbidden by the Company, was a matter of concern at the Hudson's Bay Company and commanders tried hard to show that they were honest. In Chapter 3 we discussed how commanders strove to justify their practices for the managers. While justification was infused with emotions, the accountability practices discussed here are more of a cognitive process, grounded in the building of a shared organizational memory.

As noted in the above example about the gun trade, commanders provided explanations and rationales for their actions in their detailed letters and thus built accountability. The excerpt below, where the fort commander referred to his perusal of several written documents sent to him by the headquarters, is another example of how writing supports the building of accountability:

> Gentlemen, as for explaining myself about the red and blue shirts or chequered, I cannot tell unless it be a mistake, for I have perused both the indents[6] that was sent home by the Knights and Hudson's Bay frigate and I cannot perceive by their copies the mistake, unless it be where I write

> for 6 dozen red stocking, 6 dozen blue ditto, and 6 dozen for you servants. (Letter 75, p. 305)

In spite of his efforts, he "cannot perceive by their copies the mistake" that has been made. This example shows how letters, "their copies," become an object of reference that can be shared and examined by both the writer and the managers in London. Thus, the written text becomes a key reference that allows information sharing and discussion when a different understanding or interpretation emerges. By referring to the various readings, the commander not only referred to the trace left by his "indents" (or official orders) but also to the reenactment of the meaning through each of his readings. He also referred to the reflecting process supported by writing. Moreover, as he wrote down (objectifying) the number of shirts, he recorded their numbers and also created a proof of accountability.

The commanders in their letters constantly refer to the orders given by the headquarters (objectifying)—(e.g., "And in answer to the 4th paragraph wherein your honours do direct that special care to be taken of small furs, etc.," Letter 67, p. 251)—and reassure them that they will do their best to fulfill them (e.g., "for your orders in the trading of the brass handcuffs I shall be sure to follow your instructions," Letter 4, p. 26). They reflect on their actions (reflecting) and explain in great detail (specifying) how they have enacted them (e.g., "This, may it please your honours is what always have been done, and I do promise your honours to continue so to do," Letter 67, p. 251) or why they have not followed them. For example, in the following excerpt the commander explains that he was not able to follow the instructions regarding the minerals because the miner was dead, and he refers to the previous letter in which the death was mentioned, the letter becoming a proof of his goodwill:

> As for your minerals, your miner being dead as you will find upon perusal of your former letter sent by your ship the *Knight* that I have not had an opportunity as yet, wherefore I shall desist until you shall think fit. (Letter 4, p. 26)

Hence, objectifying, reflecting, and specifying are crucial for the commanders' efforts to build accountability for the headquarters. In turn, accountability is key to achieving coordination (Okhuysen & Bechky, 2009) and thus to the building of a distributed organization. By sharing detailed information about operations and decisions, fort commanders attempted to build some trust in a mercantile organization rife with suspicions.

Providing Context

As noted earlier, the annual letters are the main means of communication between the posts and the headquarters in London. Most of the people in London have never been to the posts and therefore have very little understanding of the life conditions and the relationships between the posts' commanders, their employees, the trappers, and the Native Americans. Providing context through the addressing mechanism (i.e., the fact that the writer always writes *to someone* whose perceived needs, knowledge level, and interest he adapts the text) is another essential practice to the constitution of a distributed organization.

The annual letters sent by the post commanders to headquarters provide a simple but highly illustrative example of addressing. They always started with an acknowledgement of the reception of the letter sent by the headquarters ("Honourable Sirs, Your kind letter dated the 30th of May 1705 came safe to hand by Captain Grimington in the Hudson's Bay who thanks to God arrived here in safety the 28th of August last," [Letter 2, p. 15] and a description of the current context (e.g., weather)—"We met with a long and tedious passage to Hudson's Bay which was occasioned by contrary winds and a ship that sailed and steered badly" (Letter 3, p. 20)—so that the managers in London could imagine the situation in the faraway posts.

The post commanders also provided contextual information in order to explain why some of their orders were not fulfilled or why they took certain decisions. In the following excerpt, the commander explains to his managers why he did not send an indent with the previous ship.

> The reason why I sent no particular indent the last time I wrote was because I did not know but the goods that came over then, with what remained in the country, would have made two years trade.... For I could not divine to tell what my trade would be the summer following, but thought that after that was over, I might be the more capable to judge what goods I should want, and did not doubt to have had an opportunity to have sent home an account thereof; but the ship not returning the last fall has put me by those measures I had taken and I'm sorry this unavoidable omission should fall out this year. (Letter 1, p. 7)

In this excerpt, we see the commander reflecting on his decision and providing his readers, the management in London, with the different elements that led him to choose not to send the indent with his last letter. While justifying his actions to the management is important, writing also forces him to explain to himself the reasons of his actions. This example is similar to some of the justification and contextual information that are important in today's distributed organizations (Cramton, 2001). The addressing and specifying mechanisms are key in enabling the post commanders to share with the headquarters enough specific contextual information to allow better understanding of local decisions. One can also assume that these detailed letters are supporting the accountability that the post commanders are trying to develop. Indeed, the more details and context, the better the managers at the headquarters are likely to understand commanders' decisions and trust them.

To conclude, the letters written by the commanders and exchanged with the Hudson's Bay Company headquarters illustrate how writing is central to the creation of an organizational memory as it provides a way of recording facts about the past and the present for the future. All four mechanisms afforded by writing enable this recording process. Objectifying as a process but also as an output—documents that can be preserved and shared—is central to the recording and sharing of knowledge. The addressing and specifying mechanisms play an important role as they provide the recipients with contextual information that is key for understanding why some actions were successful and others not and why some decisions were taken. Reflecting is key to the clarification of the rationale for actions taken and thus is important for being accountable. It is important to note that while the post commanders wrote down, they did not just record what already existed, but they created new entities—new inventories, new practices. The letter writing practice was key to the enactment of an organizational memory and crucial to the performance of distributed organizing.

THEORY DEVELOPMENT THROUGH WRITING: THE CARTAN-EINSTEIN THEORY

Letters played a crucial role in knowledge creation among scientists who were often in different locations and met rarely. Writing, through its four mechanisms, enables a drafting process whereby scientists generate ideas, share them, and, based on the questions and discussions produced, modify and specify ideas and thus develop new theories. We illustrate this process with an analysis of Albert Einstein's scientific correspondence.

Developing a New Physics Theory Through the Einstein-Cartan Correspondence

Albert Einstein (1879–1955) was a world-famous German-born theoretical physicist. He is best known for the theory of relativity, and was awarded the Nobel Prize for Physics in 1921 (see Figure 4.1). He was an active and prominent member of the scientific communities of his time and developed some of his major ideas through his correspondences. Indeed, Einstein had a huge correspondence with many different recipients. He sent more than 14,500 letters and received more than 16,200 (Oliveira & Barábasi, 2005). The correspondence of Einstein with the members of his network of scientists provides an interesting parallel with today's online communities supporting scientific collaboration and idea sharing. Einstein's correspondence provides us with a rich picture of Einstein as a scientist, and as a person, and of his scientific circle. For example, the "Swiss letters" from 1902–1914 illustrate how he maintained close ties with old friends but also that he developed new relationships (particularly after 1906) with other prominent physicists such as Max von Laue and Paul Ehrenfest. During the same period, he also developed important relationships with older theorists—Max Planck, Arnold Sommerfeld, and especially H. A. Lorentz (Klein, Kox, & Schulmann, 1993).

Einstein's letters document crucial aspects of his scientific activity. Before examining this aspect more in depth, though, we need to point out that his correspondence is very complex. The correspondence between Einstein and Max Born (which we will not discuss in this book) shows the human aspect of developing science (Born, Einstein, & Born, 1971). Einstein and Born were crucial contributors to the formation of modern physics. In 1916, at the beginning of the correspondence, Einstein had just completed his papers about the general theory of relativity and was concentrating his efforts on the then-puzzling quantum phenomena. During the years that followed, Born and his students in Gottingen made

Figure 4.1 Albert Einstein during a lecture in Vienna in 1921. Photography by Ferdinand Schmutzer (1870–1928).

Source: http://hif.wikipedia.org/wiki/File:Einstein1921_by_F_Schmutzer_4.jpg

a series of discoveries on these phenomena. The correspondence between Einstein and Born illustrate their argument about the final interpretation of the quantum theory. However, their correspondence does not merely bear witness to their scientific argument; it also is a witness of the contemporary history of the years 1916 to 1954, with discussions of human, political, and ideological problems.

The correspondence with the French mathematician Élie Cartan (1929–1932), analyzed below, is a good example of the individual collaborations with mathematicians.[7] It is also a wonderful example of knowledge sharing and creation through writing, and therefore we will focus on that correspondence in this chapter. Einstein and Cartan's correspondence sets the basis for the development by the physicists Vargas and Torr (1999) of the Cartan–Einstein unification theory. Their exchange was very dense: 26 letters—not including two long mathematical notes written by Cartan—in only 12 weeks (from December 1929 to February 1930). The origins of the theory started with Einstein's attempt to transcend general relativity through the postulate of Finslerian teleparallelism. He elaborated this idea in a series of papers that started to appear in 1929. In May 1929, he received a letter from Cartan, who pointed out mathematical issues in Einstein's papers that could be solved based on some of Cartan's already published work. Einstein then

Figure 4.2 Élie Cartan.

Source: http://www.math.uni-hamburg.de/home/grothkopf/fotos/math-ges/

tried to elicit Cartan's answers to a series of mathematical questions regarding the uniqueness of the field equations he (Einstein) proposed. Einstein also wanted to know whether these equations were sufficient to determine the field fully and uniquely. Cartan gave Einstein mathematical advice on how to implement Finsler's postulate (P. Forman, 1980; Vargas & Torr, 1999). However, Einstein failed in fully building a new theory, in spite of the correct advice given by Cartan (see Figure 4.2). Later, the theory was picked up and fully developed by Vargas and Torr.

The role of writing, and specifically of letter writing (which allows a direct dialogue and challenging process), in knowledge development is illustrated in the sections below. We highlight three main practices at the core of theory development: generating ideas (i.e., writing down ideas and sharing them with others); debating ideas (i.e., engaging in a dialogue, asking for clarification, challenging the premises of an idea); and

TABLE 4.3 Development of a scientific theory through writing

Knowledge practices	Writing's mechanisms	Effects on theory development
Generating ideas	Objectifying	Draft new ideas for themselves and for the distant collaborators.
	Reflecting	Multiple readings facilitate understanding and the development of questions and suggestions for further elaboration.
Debating ideas	Objectifying	Writer and reader can refer to a specific point to discuss. Allow discussion of a specific idea.
	Addressing	Careful choice of words and syntax to create a positive dialogue.
	Reflecting	Interpret to challenge and develop the ideas.
Articulating a theory	Objectifying	Allow several drafts and changes based on feedback. Allow inclusion of complex hypotheses, formulas, or equations that could not be expressed orally.
	Addressing	Articulate the theory as clearly as possible to answer all potential questions and misinterpretations.
	Reflecting	Multiple hypotheses are discussed and then some are chosen and developed.

articulating a theory through the drafting process supported by writing. (Table 4.3 provides a summary of these three practices and the key writing mechanisms for their enactment.)

Generating Ideas

The Cartan and Einstein correspondence contains numerous ideas and hypotheses suggested by both scientists. Many of their letters illustrate the effort both made to articulate their ideas for themselves as well as for their correspondent. In the excerpt below, Cartan writes to Einstein and includes with his letter a note with a mathematical demonstration, where he tries to clarify and explain his ideas:

> I enclose a brief account of the way in which I see the question; I apologize in advance for certain details that may seem a bit irrelevant, but this is because I want to help you understand my point of view; if I have taken the wrong road, I'd like you to tell me. (Cartan to Einstein, Dec. 3, 1929)

Cartan refers to the objectifying process when he writes that he encloses "a brief account of the way in which I see the question." His vision of the problem has been objectified through writing, and he now sends it to Einstein, suggesting the possibility of a dialogue. In that sense, objectifying lays out the foundation for a drafting process for new ideas. He notes that this letter is not "fixed," but rather a first attempt to articulate his point of view ("certain details that may be a bit irrelevant") and a starting point for discussion ("If I have taken the wrong road, I'd like you to tell me"). Objectifying can be a deeply creative process that allows the expression and advancement of complex thoughts and arguments. Even once written down, these ideas are always open to refinement and discussion. Writing in that sense supports the process of knowledge creation.

Objectifying also directly supports the reflecting process. Indeed, because they offer a "materialization of the ideas," letters allow multiple readings and so facilitate understanding for the reader and the writer. Hence, Cartan writes to Einstein: "I think I can assert, having written down all possible identities and viewed the different systems likely to fit, *that there is no system MORE determined than yours*" (Dec. 22, 1929; emphasis in original). This example shows how the objectification and reflecting mechanisms helped Cartan, who read Einstein's system but also wrote down all other possible systems, to clarify and advance his thoughts and very likely Einstein's thoughts, too.

Debating Ideas

Originally, the correspondence between Einstein and Cartan started as a discussion on Cartan's contribution, but quickly the relationship became more intense. Several letters contain questions about the other's theories, each scientist trying to challenge elements of the theories they did not understand. In a letter to which he adds a technical note (which nowadays would be sent as an attachment to an e-mail), Cartan wrote:

> Cher et illustre maitre,
>
> I have received your letter and I have read and thought about it. What you say about the degree of indetermination interests me because I thought, for reasons that I outline in my note but which are a bit personal, that your system of 22 equations was perhaps *too* determined and that its solution did not have a sufficient degree of generality. But in this respect, you are infinitely more competent than I. Nevertheless, I think, that my concept of generality index, which has a precise meaning, maybe of some help. If your system, for instance depended, in the sense I indicate in my note, only on 3 or 4 arbitrary functions of 3 variables, it would certainly not be general enough.
>
> You raise objections to some of the systems I mention, objections that I am sure, are due only to the bad phrasing of my letter. (Dec. 13, 1929)

In this excerpt, where he challenges Einstein's system of 22 equations, Cartan explicitly refers to the reflecting process ("thought about it") supported by the objectifying process. Because the ideas have been objectified, they can be discussed by the recipient but also developed and referred to by the writer later, in what can be best described as an exploratory and creative process. Cartan's style and tone, his use of adverbs such as *perhaps* and *infinitely*, and his use of italics, shows his respect for Einstein and his concern not to hurt Einstein's feelings. He clearly puts a lot of effort in finding the right way to address his correspondent. Hence, he softens his questioning of the degree of generality of Einstein's system by adding, "But in this respect, you are infinitely more competent than I am." He then reiterates his challenge and adds, "Nevertheless, I think, that my concept of generality index, which has a precise meaning, maybe of some help." He also illustrates his claim with a specific example: "If your system, for instance" (specifying) and refers to a previous note (objectifying) to make sure Einstein and he refer to the same interpretation. Addressing is also present in Cartan's explanation of the objections made by Einstein: "objections that I am sure, are due only to the bad phrasing of my letter." The careful choice of adjectives and adverbs and the balanced structure of the letter (shifting between questioning Einstein's system and praising Einstein's expertise) also demonstrate how writing affords the development of subtle arguments (specifying mechanism).

Einstein and Cartan's letters illustrate a continuous and lively dialogue among correspondents. Addressing is particularly important to the enactment of this dialogue, as it leads writers to clarify their thoughts and preemptively imagine and try to answer some of the possible questions of readers. Addressing in that sense entails the specifying mechanism, as the writer develops a clearer and more specific argument to tackle all the issues that might be raised by the potential reader.

Articulating a New Theory

Cartan and Einstein's correspondence is a perfect illustration of how writing is central in the development of a new theory. Indeed, reflecting on the ideas developed and exchanged in their correspondence led Einstein and Cartan to challenge their original ideas, modify them, and then develop a new theory. Of course, this theory could have been developed in face-to-face meetings. In fact, Einstein and Cartan met several times, and these meetings played a role in the development of their ideas. For example, on December 3, 1929, Cartan referred to a previous meeting and its effects on his thought process:

> Cher et illustre maitre,
>
> It will soon be more than three weeks since you left Paris and still I have shown no sign of life. It's not that I haven't been thinking deeply about the conversation we had at Langevin's home.

He then goes into detail, explaining the solutions he has considered for the problem they are trying to solve. This example shows how the meeting (in the home of the French physicist Paul Langevin) triggered Cartan's thinking, which he then articulated and shared in writing. Einstein was able to read Cartan's complex ideas as many times as needed to understand them and use them for his own work.

The fact that writing supports the creative process can be observed in yet another excerpt from the same letter, where Cartan shares with Einstein various hypotheses and calculations triggered by their conversation at Langevin's house:

> I shall give you a few details of the theory of systems in involution which I discovered some thirty years ago that seems quite appropriate to your problem.
>
> I have also made a few calculations without being able to decide if your solution is best of all others. Another possible solution, one which contains two cosmological constants, consists in taking the 10 old equations $R_{ij} = 0$ that still have here an invariant character and adding to them 12 equations of the form. (Dec. 3, 1929)

As this excerpt demonstrates, the reflecting ("without being able to decide") and specifying ("a few details of the theory" as well as the explanation of the other possible solution) mechanisms are at the core of the exploration and drafting practices through which new knowledge is developed.

Thus, in spite of infrequent face-to-face interactions, Einstein and Cartan were able to share and develop ideas, which became the basis for a new theory, the Cartan–Einstein unification theory (Vargas & Torr, 1999). Einstein's exchanges with Cartan and the different articles they published during that period based on their exchanges represented great advances that were essential to the development of the full theory by Vargas and Torr, as illustrated by the name these later physicists gave to the theory.

Einstein's correspondence with Cartan also demonstrates the new sort of physics that Einstein began to practice (P. Forman, 1980, p. 51), where "the idea was not to solve the equations but to decide whether they possessed a solution and whether it was unique" (Wheeler, as cited in P. Forman, 1980, p. 51). In this endeavor, letters provided the right mode for a creative process in which Cartan and Einstein articulated and specified new solutions, reflected on them and their various implications, and debated them. Writing therefore allowed them to define a space of solutions rather than one single theory.

Scientific collaborations and correspondences such as the one analyzed above illustrate clearly the way writing's mechanisms support the development of new knowledge. Scientists, by writing down their ideas, are able to articulate their ideas for themselves but also for others. Once objectified, ideas can be reflected upon but also discussed, challenged, and modified. Rather than "freezing" ideas, objectifying allows writers to engage in a process of continuous change which is central to theory development. Reflecting—through which writers can deepen their own thoughts and understand better their correspondent' ideas—is undoubtedly a key mechanism for theory development. Because the writers need to address the recipient of their letters, they often end up providing more background information and specifying their ideas further. And of course, specifying is particularly important in the case of scientific formulas which would be difficult if not impossible to state orally. Einstein's correspondence with Cartan is full of long and complex equations, and it sometimes contains attached documents. Any scholar will remember the long e-mails exchanged with their coauthors, aiming to clarify the argument they were trying to develop in a paper.

WRITING AND THE CREATIVE PROCESS OF KNOWLEDGE DEVELOPMENT

Our analysis shows that writing, long considered an impoverished mode, is central to the development of knowledge both in distributed organizations and communities. Letters are a specific genre of writing whose dialogical and semiprivate nature allows us to focus on the process of knowledge development. We examined two central processes for knowledge development—knowledge sharing and knowledge creation—that have been a main focus of research in organization studies. Our analyses show that because of its mechanisms, writing is not only a recording technology, which could only support the transmission of information or explicit knowledge, but it is also as an intrinsically creative tool (Searle, 2010; Wolf, 2008).

The literature on knowledge sharing and development tends to focus on writing's recording capabilities. Thus, the study of the Hudson Bay's Company by O'Leary et al. (2002) highlights the key role played by the correspondence and by extensive record keeping in enacting trust and control and, more generally, in successful distributed working. This study provides very useful insights concerning the development of communication norms, yet the assumption is that the main capability of writing is its recording capability: By leaving a trace, it allows the creation of a shared frame of reference. Our analysis goes one step further and suggests that writing does not only record but also produces an organizational memory. By writing their orders, the committee in London created rules and procedures that should be followed. At the same time, post commanders, by writing down their actions and the rationale for them, were doing more than simply recording their doings and justifying them; they were, in fact, engaged in building the distributed organization in which they were working. Hence, Hudson's Bay Company as a distributed organization was performed through the letters exchanged by the managers in London and the post commanders in the territory of today's Canada. What our analysis proposes is a creative and dynamic perspective on the role of writing in the development of knowledge in organizational contexts.

Our research shows that some of the apparent limitations of writing—its restricted capacity for immediate feedback, the number of cues and channels available, reduced personalization (the degree to which it focuses on the recipient), and limited language variety (Daft, Lengel & Trevino, 1987; Daft & Wiginton, 1979)—can sometimes be advantages because they allow deeper reflection and increased precision, which are enacted through the objectifying act of writing, and, thus, the output is made available as a shared object that can be discussed, interpreted, and modified. Alan Newell's comment about his distance collaboration with Herbert Simon and Cliff Shaw on developing artificial intelligence is a perfect illustration of how writing can be the preferred mode of communication, as it allows the right amount of individual reflection: "Probably [...] the whole scientific enterprise with the three of us would never have worked out if we were all sitting in one place. Cliff found this way of working, with me located miles away, to be just about the right level of controlled interaction for him to flower" (McCorduck, 1979, p. 144). Interestingly, Newell suggests that not only interaction but also distance ("controlled interaction") is needed for ideas to develop and flourish. Our analysis shows that if the recording capability of writing plays an important role in the sharing and creation of knowledge, especially for the development of complex knowledge, its creative power is as remarkable. First, when we write down ideas, the ideas themselves change (Vygotsky, 1962), and thus writing is central to our capacity for abstract thinking and for developing new ideas (Olson, 1977; Wolf, 2008). Second, when we write down, we not only record what already exists but also create new entities, such as money, governments, theories, and corporations (Searle, 2010).

CHAPTER

5

Writing and Community Building

> I would like to have such a peace that we could build an Academy, not just in one city [...], but if not of all Europe, at least of the entire France, which would communicate by letters, which will be better than the talks where one gets often too excited.
>
> —Mersenne to Peiresc, July 15, 1635

The previous chapters show the power of writing in expressing emotions and in generating and communicating knowledge. So far our focus has been microscopic. We have studied correspondences and discovered the dimensions of writing that enabled correspondents who had rarely met to cultivate meaningful relationships across distances—a remarkable feat. We also discovered how writing can help produce organizational memory among physically isolated members of a vast organization and how it helps scientists develop new theories collaboratively. In this chapter, we take a more macro approach. While we continue to analyze specific correspondences, our attention is instead directed toward the liminal status of letters—how they occupy a space between the private and the public spheres—and on their role in community development and sustainment.

A lot has been said about the Internet and, more recently, Web 2.0 and social media, such as Facebook and Twitter, and how they have affected our definition of public and private spaces. Most pundits argue that these technologies allow people to share ideas and emotions as well as contribute to public debates and build communities with other people who are dispersed across geographies. Yet, while the Internet, Web 2.0, and social media have enabled many of these changes, we should not forget that the public sphere of civil society (as opposed to the state and government) had an earlier incarnation, one constituted by discussions, exchanges of letters, books, and so forth in the 17th and 18th centuries within what was called the Republic of Letters (Habermas, 1989). It was an intellectual community formed in the 17th century by *philosophes*, a term that, at the time, included philosophers, writers, and what we now call scientists. The Republic aimed to produce and disseminate knowledge across geographical boundaries, with letters at the heart of this strategy. In this chapter, we discuss how letters as a writing genre contributed to the development of this community. We focus on two periods and on two prominent members: Mersenne and Madame du Châtelet. This approach allows us to examine the role of intermediaries who were at the nodes of the correspondences among dispersed members and the way letters permitted outsiders, such as women, to be part of the community. We conclude by showing how both the liminal status of letters at the intersection of the public and private spheres, as well as writing's mechanisms, played a central part in the development and expansion of communities and in creating the public space of intellectual exchange and debate.

COMMUNICATION AND COMMUNITIES

Communication and community are intricately related. On one hand, belonging to the same community facilitates communication due to shared backgrounds, assumptions, and worldviews. Often, correspondences—and those we analyze in this book belong to this vast category—take place among people who share particular interests and even occupations (a scientist or a writer, for example). Some often met face to face, as did the Bloomsbury group, a London-based group of writers and artists to which Virginia Woolf belonged, while others met rarely, as did the Hudson's Bay Company managers and post commanders. Their membership in the same community or organization facilitated their communication.

At the same time, communication is essential to the formation and continuance of communities. Traditional communities are commonly associated with close and frequent face-to-face interactions (Tönnies, 1967; Karp, Stone, & Yoels, 1977; Putnam, 2000). For this reason, the advent of new communication technologies has spurred lively debates about the ability of technology to sustain communities. Strong virtual communities such as the WELL (Rheingold, 1993), an online community considered one of the oldest that is still functioning, which offers a wide range of emotional support and information to its members, or the open source software development communities that developed complex software products while relying almost exclusively on written exchanges, were puzzling to researchers accustomed to associating community with face-to-face communication. Earlier skepticism about the possibility of strong communal ties among dispersed individuals has given way to more nuanced attitudes that recognize that people belong concomitantly to online and offline communities and that ask whether communities can even exist without online interaction (Haythornthwaite & Kendall, 2010). As we saw, writing and letters can help build and sustain dispersed organizations such as the Hudson's Bay Company. Writing also plays an important role in the maintenance and expansion of scientific communities, as the Einstein–Cartan correspondence analyzed in Chapter 4 suggests. One reason is the mechanisms of writing; the other is the liminal status of the letters genre.

THE REPUBLIC OF LETTERS AS A PUBLIC SPACE OF EXCHANGE AND DEBATE

Never were letters more important in the development and maintenance of intellectual communities than in Europe in the 17th and 18th centuries, when the Republic of Letters flourished. According to Habermas (1989), the central event of the Enlightenment was the creation (through discussions, letters, lectures, books) of a public sphere, independent of both private and public authority (state, government). Begun as a network of private correspondence, the Republic of Letters expanded greatly, evolving into a community that spanned Europe and involved thousands of individuals. The success of this community was closely related to its ability to link dispersed scientists and philosophers and to grow continuously. While correspondences often started with a personal connection (Lux & Cook, 1998), the network gradually extended to include people who would never meet. This Republic of Letters was, in a very literal sense, formed on the basis of the letters exchanged among members, often through the intermediary of strong nodes such as Mersenne.

The main discursive practices of the Republic of Letters were "polite conversation and letter writing" (Goodman, 1994, p. 3). Face-to-face communication was important for these communities. In France, the defining social institution of the Enlightenment was the Parisian *salon*, which tended to center on academic discussion (Goodman, 1994). The British equivalent was the coffeehouse, and in Germany there were *Tischgesellschaften* (or dinner parties), all of which brought about a "sphere in which state authority was publicly monitored through informed and critical discourse by the people" (Habermas, 1989, p. xi). Other forms of face-to-face encounters were important to the development of the New Science and were enhanced by the remarkable mobility of students and professors.[1] Thus, the *peregrinatio academica* was a vast movement that encouraged students to polish their education in universities other than their own and often abroad.[2] As for famous professors, universities vied for their presence either as full-time members or as visiting scholars. For example, the French mathematician and scientist Pierre-Louis Maupertuis, who was elected director of the Academy of Science in Paris in 1742, and then admitted to the Académie Française in 1743, was then invited to Berlin and chosen president of the Prussian Royal Academy of Sciences in 1746.

Similarly, in 1641, Giovanni Domenico Cassini, an Italian astronomer at the Panzano Observatory (from 1648 to 1669) and a professor of astronomy at the University of Bologna, became the director of the Paris Observatory.

While conversation was the "governing discourse" of the salons, it was only one form of discourse. In fact, it was "the matrix within which and out of which the boundaries of the written word glowed" (Goodman, 1989, p. 136). Intellectuals often communicated through writing even when they were collocated, just as they do today via e-mail. They wrote not only to each other, but also to the public so that they could reach beyond the boundaries of the salon. Letters were the dominant form of writing, and they played a significant role in the development of the community, as we see later in the chapter. Because of its sociability and informality, "the letter-form encouraged the participation of non-specialists in scientific endeavors" (Heckendorn Cook, 1996, p. 18), allowing amateurs, women, and prominent scholars alike to participate in the vast Republic of Letters encompassing most of Europe (see sidebar The Letters Genre Within Scientific Communities for more details on how the genre of letters supported the development of distributed scientific communities).

THE LETTERS GENRE WITHIN SCIENTIFIC COMMUNITIES

The Republic of Letters studied in this chapter, just as other scientific communities, including many contemporary online communities, exhibited meritocratic values, much like Merton's idealized model of scientific community (Merton, 1942). The scientific ethos, as defined by Merton, includes four values: universalism, communism, organized skepticism, and disinterestedness.[3] These norms are held to be binding, and scientists are emotionally, as well as rationally, committed to them (McMillan & Chavis, 1986). Writing is important to achieving these Mertonian ideals of science. In particular, the objectifying mechanism supports the universalism principle, while specifying and reflecting enable the gathering and comparison of facts on whose basis organized skepticism functions.

The genre of letters, their ability to span the public and private spheres as well as their directness and sociability, was important in the New Science whereby scientists formulated limited hypotheses. In particular, objectifying and specifying allowed these individuals to record their observations and experiments and to share them with others who could repeat and confirm or deny them, as well as propose alternative hypotheses and experiments. In contrast, scientific books tended to produce universal systems (Heckendorn Cook, 1996) and were less conducive to exchanges because they were released so rarely and in small print runs (Bazerman, 1988).

> The letter encouraged the swift dissemination of discrete chunks of information such as accounts of individual experiments. Letters could be easily transmitted across national borders, escaping the kinds of censorship imposed on full-length books; they could be rapidly translated into the variety of national languages in which the New Science was now being discussed; and they could be swiftly and inexpensively printed and widely distributed. (Heckendorn Cook, 1996, pp. 17–18)[4]

These features made letters a preferred mode of communication among scientists keen to share their ideas and advances and to react to others' ideas and experiments. Of course, the danger and difficulty (not to mention the expense) of travel at the time also played a role in the popularity of letters.

Second, written exchanges have been key to scientific communities also because their members are not all collocated (Crane, 1972; Kadushin, 1966, 1968). As Kadushin has suggested, the best way to describe the social organization of a particular research area is the concept of "social circle" (Kadushin, 1966, 1968) whose boundaries are difficult to define and whose members don't all know one another. Thus, indirect interaction via intermediaries and via written communication is instrumental in the formation, sustainment, and expansion of such communities.[5]

Third, as we argued in previous chapters, letters enabled not only the sharing and development of knowledge, but also the expression and maintenance of one's commitment to the community and to the common goal of advancing science.

For all these reasons—support for scientific thinking, conduit for intermediaries who distributed information among spatially dispersed scientists, and affirmation of communal bonds—letters have been instrumental in the development and sustainment of scientific communities. The importance of writing to the development of science is such that Collins has argued that intellectual communities arose historically at the same time as public systems of writing. "What is needed is a social arrangement for writing texts of some length and distributing them to readers at a distance, an autonomous network of intellectual communication" (Collins, 1998, p. 27).

The Liminal Status of Letters: Between the Private and Public Spheres

Letters possess a particular characteristic: They straddle the two very different realms of the private and the public. Traditionally, they were seen as firmly belonging to the private sphere of intimate relations and domestic life. At the same time, letters possess a remarkable malleability that allows them to cross genres and be shared with one person or many. At one end of the spectrum are intimate letters, such as those exchanged by the soldier and his wife, which we discussed in Chapter 1; at the other end are open letters, which preserve the form of the traditional letter while being destined for a larger audience.[6] Somewhere between the personal and the open letter are gradations of how private or how public a letter can be (Beyssade & Beyssade, 1989). This liminal nature of letters has been used by many to increase the audience for their ideas. Thus, during the Republic of Letters, "the philosophers increasingly and creatively used letters to bridge the gap between the private circles in which they gathered and the public arena that they sought to shape and conquer" (Goodman, 1994, p. 137).

The liminality of letters, along with the mechanisms discussed in the previous chapters, help explain the role of letters in building and sustaining dispersed communities. In the 17th century, all members, or citizens, of the Republic of Letters were engaged in various correspondences. Furthermore, they saw it as their duty to expand the Republic through their correspondence (Goodman, 1989; Roche, 1988) because letters can be copied, circulated, read aloud in public settings, such as in the salons, and even printed and are thus accessible to an even broader audience. As letters enter the public realm, more people can participate in exchanges and debates.

The practice of *sharing received letters with others* was so well established that when writers wanted more privacy, they divided their missives in two parts: one that was destined to be shared (read aloud or copied and sent to others) and a second that was more personal (Nellen, cited in Berkvens-Stevelinck, Bots, & Haseler, 2005, p. 18). This practice had important consequences for the content of the letters. Although they were sent to a particular person, they were also composed with a more general readership in mind (Daston, 1991).

Members of the Republic shared their letters beyond their close circle of friends. As shown by Goodman (1994) in her cultural history of the French Enlightenment, members of the Republic did not produce simple letters; they "*transformed letters and correspondences into a variety of public media,* which ... retained the crucial reciprocity [that] made their readers members of a community" (p. 137; our emphasis). For example, letters were transferred to newsletters and then to journals. Thus, the earliest issues of the journal *Philosophical Transactions of the Royal Society* were largely a summary of the correspondences of Oldenburg, the secretary of the Royal Society (Bazerman, 1988). Also, "gazettes" (or newspapers) in the 18th century began as a collection of letters (Pagès, 1982). Readers were invited to respond to the ideas expressed in journals

TABLE 5.1 Writing mechanisms and community building

Writing mechanisms	Practices	Effects on community building
Objectifying	Providing shared letters that can be referred to and/or shared further.	Help create a sense of shared history.
	Articulating ideas and giving feedback.	Help create a shared understanding across different community members.
	Articulating emotions.	Identify common emotions and build relationships.
Addressing the reader	Providing contextual information (e.g., on when the letter was written or received, or about the life of the reader).	Facilitate understanding and a sense of shared context, which helps create a sense of *we*-ness.
	Referring to the community as a persona.	Performing and creating identity by the reference to the community as a person.
	Providing background information and rationale for a shared idea.	Help create a set of shared references and facilitate the development of knowledge.
	Providing extra information to clarify an emotional reaction that might be misunderstood.	Help create a relationship and develop trust.
Reflecting	Referring to the writing or the reading of the letter—single or multiple readings.	Help create a dialogue with the reader and support the development of a relationship.
	Referring to the interpretative process.	Provide a sense of commitment and potentially create trust.
	Sharing impressions and emotions about the community, its members, and its evolution.	Support a sense of belonging and an awareness of a shared identity.
Specifying	Long descriptions.	Allow writers to develop a shared understanding and thus support relationship development.
	Providing a lot of details and/or nuance.	Allow writers to have a sense of shared context and thus the development of a sense of *we*-ness.

Note. In this table we focus only on the effects of written exchanges on the development of a sense of common belongingness, of "*we*-ness" among correspondents. At the same time, because the bonds among the members of the Republic of Letters were resting heavily on the expression of emotions and on the development and sharing of knowledge, the development of the community relies on some of the practices we have examined in the previous chapters.

and periodicals, which published their letters, thus introducing them to the community of those journals or periodicals. By inviting readers to engage with them and react to their work, 18th-century writers created their own public, while simultaneously writing for it (Goodman, 1989). Hence, the liminal nature of letters allowed them not only to support relationships among members of the Republic through sustained epistolary relationships, such as those developed by Mersenne with his correspondents or Madame du Châtelet with hers, but they also allowed the engagement of a broader audience through shared letters. This engagement gave birth to other media, such as journals and gazettes.

We now turn to a close examination of the way in which the liminal status of letters helped sustain and expand the community through the actions of two types of members and their reliance on writing and its mechanisms. The first is intermediaries who reached out to current and potential members by disseminating, linking, and forwarding letters. The second is outsiders, such as women, whose writing allowed them to reach into existing networks and communities. Table 5.1 presents an overview of the writing practices used by members of the community such as Mersenne and Madame du Châtelet and how they helped develop a sense of community among its numerous dispersed members.

THE USE OF WRITING BY COMMUNITY INTERMEDIARIES: MERSENNE AND THE REPUBLIC OF LETTERS

At a time when there were no academic journals and few scholarly institutions, the expansion and success of the Republic of Letters rested largely on an intricate system of intermediaries and their practices. It is probably fair to say that most community members played, at least to some extent, the role of intermediary. Others, such as the prominent Erasmus and Leibniz, acted as central nodes in extensive networks throughout Europe.

One such central node was Marin Mersenne (1588–1648), a Jesuit priest and scholar who lived in a Parisian monastery when he was not traveling elsewhere for his studies; from 1629 to 1630, for example, he studied in Holland. His impressive correspondences (17 volumes) with the main intellectuals of his epoch led contemporaries, such as Thomas Hobbes, to consider him "the axis around which planets were revolving" (Bots & Waquet, 1997, p. 76). The network of Mersenne's correspondents reveals the importance of letters in building and sustaining an intellectual community that transcended face-to-face contexts. Although centered primarily in Paris, and in France, more generally, Mersenne's network dominated intellectual exchanges in Europe in the second quarter of the 17th century. He is considered a cultural intermediary who helped cut across the national and religious boundaries that traditionally divided Europe (Berkvens-Stevelinck, 2005).

A Network of Correspondences

The 1,135 letters that composed the 17 volumes of Mersenne's preserved correspondence contain 330 written by Mersenne and 805 written by approximately 100 respondents. Mathematicians, medical doctors, astronomers, physicists, and philosophers wrote more than 60% of them. While almost half of the letters Mersenne received came from France, the others were from correspondents scattered throughout Europe. He maintained valuable connections with many of the most important scholars of his time (e.g., the philosopher Descartes, the mathematicians Pierre de Fermat and Evangelista Torricelli, the poet Constantin Huygens, the theologian André Rivet). Figure 5.1 provides a visualization of his correspondence network, emphasizing its richness and the significant role a core group of 13 correspondents played (50% of the letters were written by them). This figure also illustrates how Mersenne, as Hobbes noted, sat at the center of a scientific community that addressed astronomy, moral philosophy, and science.

Mersenne's main contribution to his time was to popularize among his contemporaries groundbreaking philosophies, especially those of Descartes and Galileo. From the beginning, he wanted to make those philosophies accessible to a larger public. He also thought that the advancement of science required collective work. To achieve this goal, he endeavored to create an academy of scientists much larger than his small circle of friends who were passionate about mathematics. In a letter to Peiresc,[7] dated July 15, 1635, he expresses this idea of developing a larger community and the role of writing in this enterprise:

> I would like to have such a peace that we could build an Academy, not just in one city [...], but if not of all Europe, at least of the entire France, which would communicate by letters, which will be better than the talks where one gets often too excited. (cited in Bots, 2005, p. 175)[8]

In pursuing this goal, from the 1620s onward, Mersenne sought a community of scholars among whom political, religious, and scientific differences were inconsequential. He was quite successful. As one of his contemporaries wrote:

> He had become the center of the world of letters, owing to the contact he maintained with all, and all with him ... serving a function, in the Republic of Letters similar to that of the heart in the circulation of blood within the human body. (quoted in Goodman, 1994, p. 20)

In 1635 Mersenne became the initiator of *Academia Parisiensis,* which brought together scholars such as Blaise Pascal and his father, Etienne, and mathematicians Claude Mydorge, Claude Hardy, Gilles Personne de Roberval, and Pierre de Fermat. This academy was informal and lacked status, yet Mersenne's network formed the basis on which the French Academy of Sciences and the English Royal Academy would be built (Collins, 1998).

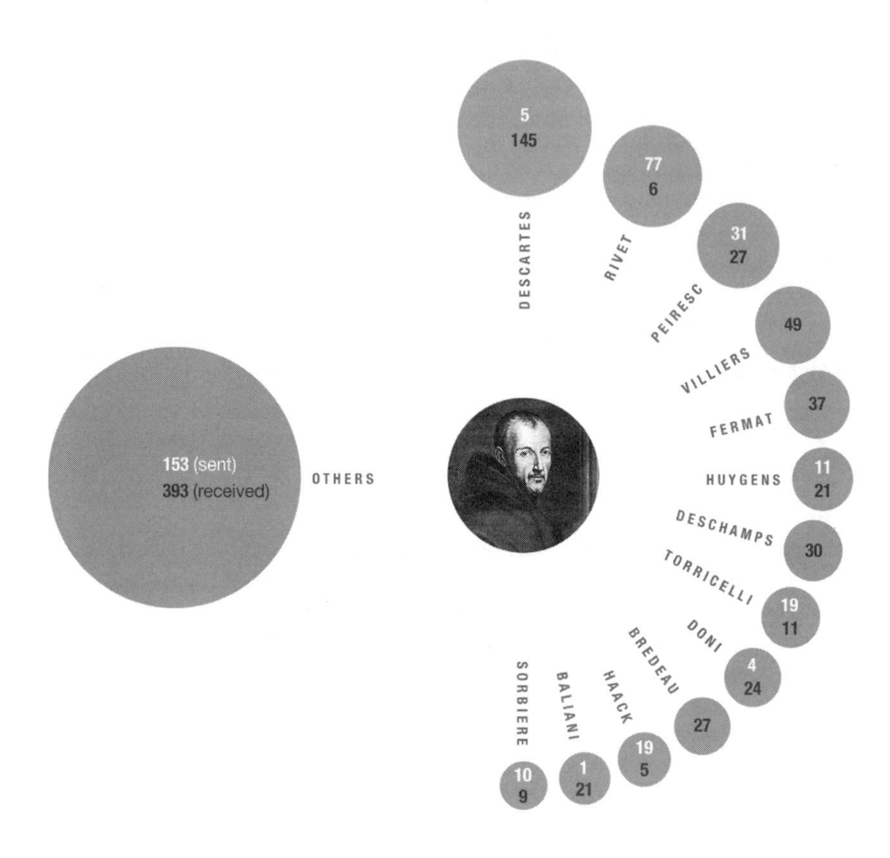

Figure 5.1 Infographics of Mersenne's network. © 2010 by Guilhem Tamisier. The analysis was based on the data provided by Hans Bots (2005) in "Martin Mersenne, 'secretaire general' de la Republique des Lettres (1620–1648)," in Berkvens-Stevelinck, Bots, and Haseler (2005).

Source: http://simple.wikipedia.org/wiki/File:Marin_mersenne.jpg

In this way, Mersenne's "republic of letters" provided the organizational hub for generations of modern Western philosophers (Collins, 1998).

As Figure 5.1 shows, Mersenne seems to have exchanged the most letters with Descartes, but their relationship was asymmetric. While

Descartes wrote 145 letters to Mersenne, only 5 letters of Mersenne to Descartes have been found. This imbalance might reflect the student-master relationship, as Mersenne had been one of Descartes's high school professors. It could also be that not all of letters sent by Mersenne have been found. Finally, it may be that Descartes, like other scholars of the time, was worried about being troubled by the Church and, thus, did not publish many books, but rather sent letters (Bridoux, 1953). Because of their liminal status, letters could be kept private and thus protect their writer from the dangers of censorship while still being shared with a select group of trusted scholars through nodes such as Mersenne.

Mersenne and Descartes's Epistolary Relationship: Developing Ideas and Building the Community

The correspondence between Mersenne and Descartes shows the role of the exchanges in the development of Descartes's ideas. Part of the explanation rests with the mechanisms of writing, just as we saw in the previous chapter where we analyzed the Einstein-Cartan letters. Through the objectifying mechanism, Descartes developed his ideas and shared them with Mersenne, who read and reflected on them. In his letters, Mersenne asked Descartes questions based on his own interpretations and thus pushed his former student to develop his own theories. For example, their correspondence was crucial to Descartes's articulation of his first metaphysical thesis (on the creation of eternal truths) in three letters written in the spring of 1630 (Beyssade & Beyssade, 1989).

The beginning of these letters illustrates the addressing mechanism, with Descartes providing contextual information to Mersenne and discussing more personal and emotional matters. The latter inclusion positioned the letter in the private sphere. Descartes then refers to the query raised by Mersenne:

> Regarding your question, whether we can explain the cause of the beautiful, it is similar to what you asked me previously, that is, why a sound is more agreeable than another, except that beautiful seems to refer to the sight rather than the audition. (Descartes, 1953b, March 18, 1630)

The quote also illustrates Descartes's reflective process—"it is similar to what you asked me previously"—and he builds on his argument by developing the sound example. All these letters are detailed and articulated, thus illustrating the specifying mechanism and the fact that some philosophical arguments cannot be developed unless written (Searle, 2010). For instance, in these three letters, Descartes states that God freely created the truths that we call eternal. Cartesian scholars refer to this aphorism as "the doctrine of 1630" or "the creation of eternal truth" (see, for example, Devillairs, 2001), and argue that Descartes's later work—the *Principle of the Philosophy*—builds on it.

Figure 5.2 *Portrait of René Descartes* (1596–1650), oil on canvas, reproduction by André Hatala [e.a.] (1997) De eeuw van Rembrandt, Bruxelles: Crédit communal de Belgique, ISBN 2–908388–32–4.

Source: http://en.wikipedia.org/wiki/File:Frans_Hals_-_Portret_van_Ren%C3%A9_Descartes. jpg

The exchange between Mersenne and Descartes illustrates the key role of letters in the development of Descartes's theory: not only the impact of Mersenne's comments on the theory development but also the influence of letters due to their liminal status. Indeed, Mersenne was instrumental in disseminating Descartes's work to some of the most prominent scholars at the time, such as Antoine Arnauld (1612–1694), Pierre Gassendi (1592–1655), and Thomas Hobbes (1588–1679). For example, he sent these scholars and other theologians and philosophers *The Meditations*, one of Descartes's major works. Mersenne shared with Descartes his correspondents' critiques and objections, and Descartes replied with letters, which were published at the end of the second edition of *The Meditations*. The importance of these exchanges is demonstrated by the fact that today all scholarly editions of *The Meditations* include the objections and replies. Through his voluminous correspondence with Mersenne, and regardless of his actual physical location (Descartes spent the last 20 years of his life in Holland, and traveled extensively in Europe before that), he remained connected to numerous intellectuals who provided meaningful commentary on his work.

Intermediaries such as Mersenne did more than simply forward letters (or parts thereof) among their contacts. Because letters could be both private—including bits of personal information, political news, and gossip—as well as public (they could discuss scientific principles for later publication, for instance), intermediaries edited them to avoid the censorship at play in a Europe largely ruled by despots. For example, Henry Oldenburg "diplomatically pruned letters of 'all Personal Reflections,' polished compliments as well as rude remarks, before publishing them in the *Philosophical Transactions*" (Shapin, quoted in Daston, 1991, p. 420). Such practices had powerful consequences for the formation of communities because intermediaries used the letters' liminal status to the community's advantage.

Thus, letter exchanges were instrumental in creating, maintaining, and expanding the scientific community of the 16th and 17th centuries. As discussed in Chapter 4 and illustrated in the letters exchanged between Descartes and Mersenne in the spring of 1630, the mechanisms of writing supported the knowledge creation and sharing process. Moreover, intermediaries used the letters' liminal status to involve relevant experts in scientific exchanges and to expand the community to include faraway individuals and a larger public.

The same two traits of letters—its writing mechanisms (objectifying, specifying), as well as its intermixing of the private and public spheres—nourished scientific communities in other ways as well. For example, through the participation of outsiders such as women, who could not officially become members of scientific communities, letters expanded the circle of scientists. In the following section, we analyze more closely the correspondence of Madame du Châtelet, one of the few women who participated in the intense network of scientific and intellectual exchanges that constituted the Republic of Letters.

THE USE OF WRITING BY OUTSIDERS TO GAIN ACCESS TO COMMUNITY EXCHANGES: MADAME DU CHÂTELET AND THE REPUBLIC OF LETTERS

Women occupied a particular position in the Republic of Letters. Formal forums for knowledge acquisition and development, such as universities, were closed to them. Although the statutes of the Academy of Sciences did not bar their admission, a firmly established tradition of exclusion from the prestigious institution was in place (Petrovich, 1999). Thus, from its creation in 1666 until 1979, the Paris Academy of Sciences did not include women as full members. Also, for many years universities did not accept women students or professors.[9] Even as science was becoming professionalized, women were confined to an amateur status, despite their desires to practice in the field (Gardiner, 1984).

Their lack of formal education was not always a barrier, though. Some women who belonged to the social elite in France and Italy participated in the political, cultural, and scientific exchanges of their time (Dalton, 2003). Some belonged to a public sphere of government critique made up of salons and the press (Goodman, 1994), while others fostered exchanges among scientists and philosophers by hosting salons. The salons of Madame Geoffrin (1699–1777) and Julie de Lespinasse (1732–1776), for example, entertained the Encyclopedists[10] and philosophers, such as David Hume.

Several women did make significant contributions to science, however. Madame du Pierry (1746–1789) became the first woman professor of astronomy in Paris; Hortense Lepaute (1723–1788) made original calculations regarding the oscillations of pendulums of varying lengths and was hired by prominent mathematicians of the time to calculate the attraction of Jupiter and Saturn on Halley's Comet; Madame Lavoisier (1758–1836—the wife of Antoine-Laurent de Lavoisier, considered the father of modern chemistry) assisted her husband and finished his work after his death (Gardiner, 1984). But in spite of their accomplishments, there is scant information on these women's work and lives. Du Châtelet is one of the few about whose work and life detailed information has been preserved.

Émilie du Châtelet's Correspondence

Gabrielle Émilie Le Tonnelier de Breteuil, Marquise du Châtelet, known as Émilie du Châtelet (1706–1749) was one of the first women scientists in Europe, a physicist, mathematician, translator, and essayist whose greatest work was to translate from Latin to French, and to comment on, Newton's *Principia*. Published in 1759, du Châtelet's is the authoritative French translation of Newton's work because she not only translated, but also

> used the simpler language of calculus to express Newton's proofs. She also articulated the implications of Newton's work for the understanding of gravity and energy, which were essential to further developments including Einstein's $E = mc^2$. (Bodanis, 2001).[11]

Du Châtelet's other accomplishments included *Institutions de Physique*, published in 1740, which explained the metaphysical theories of Leibniz found in his *Monadologie* (1714). By Voltaire's own account, du Châtelet also worked with him on *Elements of the Philosophy of Newton*, although

Voltaire is listed as its sole author. Finally, du Châtelet translated from Greek into French a play by Sophocles. The fame she gained, especially through the *Institutions,* was unusual for a woman in the 17th century.

But it is her correspondence that best demonstrates her centrality among Newtonians and her contributions to developments in both the Paris and Berlin academies (Hutton, 2004). Having earned her expert status through the *Institutions,* du Châtelet's letters became what the European scientific community turned to when gauging the hot topics of the day—either du Châtelet's own work or what she thought was important (Bodanis, 2007).

Du Châtelet had a long-lasting relationship with Voltaire (1694–1778), who considered her his soul mate. Voltaire was already very famous when they met, but du Châtelet's own renown grew after the publication of her books and after her essay on fire was ranked number six at the Academy of Science's competition in April 1738—many researchers acknowledged her contribution was best (Bodanis, 2007). Together, she and Voltaire created a research institute in Cirey, an isolated chateau in the French countryside, where they worked and collaborated and where many European intellectuals visited. The couple and their guests came to form an important node in the scientific networks of their time. Indeed, "people who received letters from Cirey accordingly tended to share them: sometimes with just a few closed friends, but often formally recopying them, to send to even greater number of contacts. It was much like a private Internet: with information and opinions being fed in, then swirling from one node to another across Europe" (Bodanis, 2007, p. 114).

Cirey was not the only way the duo affected scientific circles of their day. Their letters were also influential. Like many scientists of her time, du Châtelet was an inveterate and varied letter writer. Thousands of her letters have survived, including those she exchanged with Voltaire, houseguests, neighbors, purchasing agents, and scientists. Her private correspondence included letters to confidantes, such as Abbé de Sade and Richelieu; love letters (to Saint-Lambert after 1748); business letters; society missives (letters, or "billets") in which she offered and asked for news, or talked about others and about Voltaire (Bonnel, 2000).

The flurry of letters and texts that poured out from Cirey was important for the scientific ideas they contained. They also inspired young thinkers, such as Helvetius, d'Alembert, and Diderot, whose work further spread Enlightenment ideas. Letters also arrived from important scientists in many European countries. Du Châtelet and Voltaire would share the mail over breakfast and, for example, read letters from Bernoulli, Frederik the Great, Bolingbroke, or Jonathan Swift.

So, despite her exclusion from official, all-male institutions, du Châtelet, through her publications, but primarily through her letters, became "a respected correspondent with top researchers in England, Italy, and France" (Bodanis, 2007, pp. 183–184) and came to be involved in the

Figure 5.3 *Portrait of Émilie du Châtelet* (1706–1749), by Maurice Quentin de La Tour 1704–1788.

Source: http://en.wikipedia.org/wiki/File:%C3%89milie_du_Ch%C3%A2telet_1.jpg

discussion of many scientific developments of the time. In a letter to her tutor, Maupertuis, she expressed strongly and movingly her reflections and feelings of dissatisfaction about her inability to pursue her true calling—science—and her desire to interact closely with other scientists:

> Life is so short, so full of pointless duties and details, that having a family and a house, I don't deviate at all from my little study plan to read new books. I despair at my ignorance. If I were a man, I would be at Mont-Valérien [Maupertius's residence] with you, and I would abandon there all of life's futilities. I love the study more passionately than I loved the world; but I realized it too late. Your friendship is a comfort to my ego. (Letter XXVIII, Oct. 24, 1738)[12]

Figure 5.4 *Portrait of François-Marie Arouet de Voltaire* (1694–1778), oil on canvas, by Catherine Lusurier (around 1753–1781) inspired by Nicolas de Largillière (1656–1746); 1778 (copy), 1718 (original).

Source: http://en.wikipedia.org/wiki/File:Voltaire.jpg

As suggested in this excerpt, it is through letters that du Châtelet was able to be involved in the scientific community of her time.[13] She exchanged letters, for example, with nine prominent scientists, such as the mathematicians Jean Bernoulli and Clairaut (Bonnel, 2000), as well as Father Jacquier, a famous mathematician whom she met once at Cirey and who helped her get admitted to the Bologna Institute (Badinter, 2006).

Figure 5.5 *Portrait of Pierre Louis Maupertuis*, 18th-century drawing.
Source: http://en.wikipedia.org/wiki/File:PierreLouisMaupertuis.jpg

Many of the letters were written to Moreau de Maupertuis (1698–1759), who was a member of the Academy of Sciences and became du Châtelet's geometry tutor. He was also a respected mathematician, astronomer, and physicist, and he supported Newton's hotly debated theories. Maupertuis also participated in one of the first scientific expeditions when Louis XV sent him and others to Lapland to measure the length of a degree of arc of the meridian. Du Châtelet's correspondence with Maupertuis illustrates the diversity of letters that can exist even between two correspondents—from the short note to set up an appointment ("come today at 6 o'clock"[14]) to the letter describing everyday life ("I went back to geometry these days"[15]) and to those that gossiped ("Regarding Monsieur Clairaut, I believe he went back to Madame de Theil: he broke his word to me Monday and Tuesday"[16]), as well as those that expressed matters of the heart (the romantic letter) and those that expressed matters of the mind (scientific missives that tested suppositions).

Their correspondence illustrates how letters, as discussed in Chapters 3 and 4, were central to the generation of knowledge as well as to the expression of emotions, both constitutive elements of the scientific community to which Madame du Châtelet belonged. Furthermore, these letters support the enactment and expansion of the community by allowing a space in which outsiders could participate. The letters' liminal status played a triple role in this respect. First, letters established an exploratory space where writers could examine their thoughts, play with ideas, and get feedback from others. Second, their public nature provided key proof of scientific contributions and ownership. Last, letters were often shared with third parties, a practice that enhanced and maintained the community.

Letters as a Space for Experimentation

For those who could not participate in academy meetings and for those who had to dress up as a man just to enter the coffeehouse where scholars gathered after official meetings (Bodanis, 2007), correspondences were a life jacket. Du Châtelet's extensive communications with Voltaire, Maupertuis, and others served as a laboratory for her ideas, hypotheses, and theories (Bonnel, 2000). The letters to Maupertuis from the summer of 1738 are part of the letters-as-lab (where she did some thought experiments) that helped her write the *Institutions.* In a letter from June 21 (i.129), she defines the research topic: "Metaphysical reason prefers the attraction law that follows nature." Then, on July 17 (i.133), she articulated the question and hypothesis, as this excerpt illustrates:

> If atoms are matter, which they must certainly be as a composite cannot be made of parts completely different from itself, if, I say, they are matter, then they must have a shape. (Bonnel, 2000, p. 93)

She ended the letter by asking Maupertuis to identify the missing assumption and confirm or contest the demonstration, thus asking him to reflect on her ideas. The letter of September 1 (i.139) notes that Maupertuis agrees with the demonstration. In the same letter, she criticized an argument made by the mathematician Mairan in a recent treaty (objectification) and explained why she thought Mairan used "invalid reasoning."

As these examples demonstrate, letters communicate thoughts, but they also provide a research space, a place for thinking aloud. In this type of letter, the writer's reflections are shared with the recipient. Before making her ideas public, she tests them in letters. The recipient thus has time to read and understand. Objectifying is a key mechanism here, as it allows the reader to go back and even refer to the specific idea and ask for clarification. For example, in a letter to Maupertuis, du Châtelet quoted a passage in a previous letter from him and requested clarification:

> In one of your letters about springs, you say that all the physical explanations of the spring are so poor that you would rather believe that when two bodies meet, and have then the tendency to move away at the same speed as the one with which they came close, that they are subject to the general law of joy; I suppose joy is there for another word, that I wasn't able to provide, and which I beg you to clarify. (Letter XXXIII, Dec. 20, 1739)

Du Châtelet addressed the letter specifically to her correspondent (addressing mechanism), requesting approval or criticisms of her ideas. Thus, through the use of the reflecting and specifying mechanisms, the letter becomes a kind of draft. Other examples from her scientific letters trace du Châtelet's progress through explicit musings ("What would happen if?") and daydreams ("I imagine…") (Bonnel, 2000). She sometimes guesses questions her correspondent might pose and replies to them in a letter. The semiprivate space of letters means she can make conditional queries and explore gut reactions, both lines of inquiry forbidden in published treatises, which only allow proven demonstrations (Bonnel, 2000).

Thus, the letters-as-lab are where Madame du Châtelet criticizes, asks for advice and feedback, tests her intuitions, and affirms her theories. They also offer a private space of reflection, as well as a dialogue with others of like mind. In the private sphere, the letters to Maupertuis show her evolution as her knowledge grows: from pupil to knowledgeable amateur to colleague (Bonnel, 2000). At the same time, they are a step toward her participation in the public sphere of knowledge exchange.

Letters as Proofs of Scientific Contributions

One consequence of the objectifying mechanism is that letters can be used as proofs of one's ideas. When a written document, originally private, is shared with the wider public, it can attest to an idea's origin. Thus, when a controversy arose around an idea put forth by Monsieur de Mairan, the permanent secretary of the Académie Royale des Sciences, du Châtelet turned to her correspondence for proof that she was the first person to criticize his thesis.

The controversy arose while she was working on the first edition of *Institutions Physiques*, published anonymously in 1741. While working on the book, du Châtelet asked the mathematician Samuel Koenig for some summaries. According to her, Koenig spread the rumor that her book was worthless and that he had written another one for her (Bonnel, 2000; Hutton, 2004). Following this incident, on March 22, 1741, she wrote to Maupertuis:

> You are the only one who knows whether it is Monsieur De Koenig or I who critiqued Monsieur De Mairan's dissertation, because I wrote to you at St. Malo in 1738, long before I knew Koenig even existed, almost the same things about it that are in my book. (Bonnel, 2000, p. 94)

Letters, through the trace left by the written word, provided solid proof of one's thoughts. The same objectifying mechanism was at work in Voltaire's intervention to clarify the controversy. Voltaire wrote to Mairan: "I have the written proof of what I tell you. Her beliefs started to falter before she had met the apostle of monads [Koenig] who corrupted her and before having seen Jean Bernoulli the son" (Badinter, 2006, p. 344). By sharing her letters with Voltaire, du Châtelet sought the help of prominent community members to defend herself against false accusations. Again, through written proof, Voltaire can support du Châtelet's honor in this particular dispute.

To prove that she was the first to have criticized Mairan's ideas, in the 1742 edition of the *Institutions Physiques,* she went even further in proving her primacy to the widest audience. In this new edition, she published two letters: the one written by Mairan after the first edition, and the response she sent him on March 26, 1741 (Hutton, 2004). By moving these letters from the private to the public sphere—and of course by publishing her book—Madame du Châtelet could prove to the world that she was the first to have criticized Mairan's ideas. According to science historians, "The printing of the correspondence with Mairan in the 1742 edition has all the hallmarks of an affidavit confirming the competence and independence of the author in matters of physics" (Hutton, 2004, p. 527).

Letters as Conveying Information and Supporting Relationships

Earlier we saw Madame du Châtelet at the center of a major correspondence network of the Republic of Letters, receiving from numerous philosophers and scientists letters she discussed with Voltaire and other Cirey guests. Like other participants in the Republic of Letters, du Châtelet was an intermediary in this community. In this section, we want to highlight two roles played by Madame du Châtelet (and played by the great intermediary Mersenne as well): that of trusted liaison for close friends and that of information conduit. Indeed, letters not only transmitted information, news, ideas, and gossip, but as we discussed in Chapter 3, they also expressed emotions: "They also embodied the writer and thus, on another level, functioned as a substitute for conversation, an absence made presence" (Goodman, 1994, p. 143). Hence, du Châtelet could write to Maupertuis, "I hope that the first post will bring me news from you: only your letters can stand for the elegance of your imagination and spirit" (Letter VI).

These roles—trusted liaison and information conduit—are best understood by the notion of letters as a *relation* (Dumonceaux, 1983;

Goodman, 1994). The letter is a relation at two levels. First, it represents both correspondents and the relationship binding them. Second, as suggested by Pierre Dumonceaux, through its content, it "relates" (gives an account) and so provides food for thought and future discussions, either in letters or face-to-face encounters. Thus, correspondences help build large communities among collocated individuals, as well as among those who have never seen one another and who may not even know of the other's existence.

One person for whom du Châtelet acted as trusted liaison consistently was Voltaire. As his companion for many years, she was one of few who knew where he was during his long, recurring exile periods, and it was she who played the role of intermediary for those extended durations. For instance, she enclosed in a letter to Voltaire dated May 22, 1734, one of Maupertuis's letters to him, as Voltaire was, at the time, once again in exile. She then wrote to Maupertuis:

> It has been a long while since I wrote, Monsieur, and since I received news from you; I was saddened that you sent me two copies of "Le Pour et Contre" in the same envelope, and thus stole me a letter.
>
> I forwarded yours to Voltaire; if it reaches him, I know he will be extremely pleased; I completely ignore his fate, I haven't had news from him since he left. (Letter IX, May 22, 1734)

This example shows how important letters from Maupertuis were for du Châtelet and how she kept Voltaire in touch with Maupertuis even when such linkages were difficult to make. This example also shows how, through her correspondence, du Châtelet kept abreast of the latest scientific developments: the book *Le Pour et Contre* was to be published by Abbé Jean-François Prévost one year later. Through the use of the addressing mechanism, du Châtelet refers to the fact that it has been a long time since she last wrote her former tutor and since she had received news from him. Of course her reference may be formulaic, a routine politeness. Still, her entire correspondence with Maupertuis displays a constant and careful attempt to adapt her tone to the stage of the relationship, to remind him of and thank him for his role in her learning, to praise his accomplishments, and, ultimately, to maintain the relationship.

Forwarding was not the only means in which letter exchanges kept members informed about one another's work. Many letters conveyed information about other correspondents' ideas, thus disseminating them among other scientists. Moreover, in some cases, when someone learned of a theory proposed by a scientist whom they might not know personally, or at least not very well, they asked another member whom they knew better to be an intermediary. For example, on November 19 and December 1, 1738, Madame du Châtelet shared with Maupertuis her skepticism at the possibility that Du Fay, a French scientist,[17] proved that there were three primary colors and not four, as Newton had demonstrated:

> I know almost by heart M. Newton's optics and I confess that I didn't think one could dismiss his experiences on refrangibility. If M. Du Fay has strong reasons I am ready to hear them.... A serious series of experiments are needed to undermine a truth that M. Newton has examined thoroughly. For the moment, as I haven't seen any of Du Fay's experiences, I suspend all judgment. (Bonnel, 2000, p. 81)

The example illustrates the feverishness with which the community discussed and sought to learn about possible new scientific developments and experiments. It further suggests that du Châtelet would like to learn, from Maupertuis or other scientists up to date on the issue, if there were any experiments to support Du Fay's claims. However, her curiosity pushes her to also write directly to Du Fay. As soon as he answered that he had no experimental proofs for his ideas, she wrote again to Maupertuis:

> I received a very wise letter from Du Fay about his new ideas about colors; I'm afraid he might be forced to abandon them; but I am pleased to see, from his letter, that he will relent if he doesn't have experimental support. (Letter XXX, Dec. 28, 1738, i.159)

In this letter, Madame du Châtelet conveys information about Du Fay's latest work, but beyond simply sharing the information, she also provides critical commentary about the work—"I think he will have to abandon them"—thus engaging in the scientific conversation of the time and possibly influencing it. Later in the letter, she refers to a visit Du Fay made to Cirey and the fact that she gave Du Fay several prose and verse texts to pass on to La Condamine,[18] a French naturalist and mathematician both Maupertuis and du Châtelet knew. This last note shows du Châtelet again disseminating ideas and being a central node of her scientific network.

The example also shows that du Châtelet, as did other scientists within the Republic of Letters, reasoned according to the principles of skepticism and universalism (Merton, 1968) and therefore approved a claim on the basis of accepted criteria and solid empirical evidence. Hence, she assessed Du Fay's work on the principle of experimentation—"if he doesn't have experimental support"—and assumed that the absence of experimental support would lead Du Fay to abandon his theory. It also shows how quickly she conveyed the scientific news (or lack thereof, in this particular case) to others interested in the same matter.

This type of event was by no means singular. The liminal space letters occupied regularly sustained intense debates within the sciences. Du Châtelet and Voltaire often read each other's scientific letters. For example, she wrote Maupertuis that what he told Voltaire in his letter about circular movement was proof of what she told him a few weeks before (Jan. 20, 1739, i.175). Also, on February 27, 1739, she replied to a question that Frederic the Great (the King of Prussia) had asked Voltaire. Still the next day, Voltaire wrote to Frederic himself on the same topic. The intensity of these epistolary exchanges is similar to what happens today in some online forums. In fact, Bodanis (2007), who authored a biography of Madame du Châtelet, emphasizes that sometimes she sent "four or five messages a day, and Voltaire wrote constantly too, so what I found about key events was often as dense as an email trail today" (p. 5).

Our examination of du Châtelet's correspondence shows how letters allowed her to participate in the intellectual communities of the time and thus fulfill her needs to acquire and share knowledge. Letters' liminality made it possible for outsiders, such as women, to participate and thus to expand the community. Women could engage in private letter exchanges first and so not "threaten" the established order. Through her letters, du Châtelet became engaged in a continuous discussion of Newtonianism and other theories with the scientists of her time. Furthermore, her letters could be made public and so could prove her expertise. Thus, her letters created a space for her participation in domains, such as science and philosophy, forbidden to women. In addition, her correspondence helped sustain the scientific community to which she belonged.

The thriving community of the Republic of Letters, whose lasting effects include the very foundations of contemporary approaches to science, bears striking similarities—the mixing of the private and public, the intensity of the exchanges, the continuous influx of newcomers and contributors—with current meritocracies that function mostly online, such as the open source software development communities.

PART 1: THE RELEVANCE OF THE MODE OF COMMUNICATION PERSPECTIVE

Our analysis of letters shows that writing, long considered an impoverished mode of communication, permits complex ideas to be shared and subtle emotions to be expressed in intellectual communities as well as in distributed organizations. This discovery suggests that focusing on the writing mode rather than the media allows us to explain central phenomena in organizations, such as the sharing and development of new knowledge, the expression of emotions, and the expansion of dispersed communities. The portrait of writing that emerges from these chapters is that of a rich mode that has enabled tremendous accomplishments.

A main goal of our work so far has been the elaboration of the four mechanisms of writing—objectifying, addressing, specifying, and reflecting—that result from our analysis of correspondences in various contexts. We uncover the explanatory power of the mechanisms of writing by examining how letter writers exchange and express emotions, develop knowledge, and build communities. Media richness and social presence theorists focus on the amount of information transmitted and claim that, because oral communication relies on several channels (verbal and nonverbal), it allows partners to express themselves fully and hence form a sense of one another. They also argue that written communication is disembodied and inflexible, involving mostly simple and straightforward messages. In contrast, our analysis shows how the other is, in fact, never absent in written communication and that written communication can support the sharing and development of complex ideas as well as the expression of nuanced emotions. Moreover, while media theories predict oral communication to be more efficient in negotiating interpretations, solving complex problems, and managing emotions (because of its multichannel and responsive characteristics), we show how writing objectifies emotions and ideas, thus affording the writer and the reader time to reflect and analyze. These characteristics are essential in the pursuit of nuanced emotional expression and of idea development, which are the building blocks for collaboration, for nourishing relationships between people at a distance, and for the development of communities.

Our analysis takes up the rich tradition of historical studies of organizational communication (Bazerman, 1988; O'Leary et al., 2002; Yates, 1989) and demonstrates the usefulness of a historical perspective in developing a deep understanding of contemporary phenomena. Documenting past communicative practices in letter writing helped us to better understand the processes underlying current communication practices.

The mode perspective advanced in the first part of this book constitutes a general theory of communication that centers on the ways in which the mode of communication—written, oral—affects the relationship between sender and recipient. It can therefore be used to study a wide range of communication media, from personal letters that have been important historically to the newest forms of communication. Moreover, because it emphasizes the underlying (and constant) dimensions of the modes of communication, our focus on the mode rather than the medium is better suited to the evaluation of the rapid rate at which communication technologies evolve and to understanding their implications for organizational communication.

In Part 2, we illustrate the explanatory power of our perspective in the context of new media and online interactions. In the following chapters, we show how the mechanisms of writing can produce a fruitful analysis of modern interactions via e-mail, blogs, public forums, and other online media. We also delineate for organizational and communication scholars some avenues for future research that looks at the mode of communication, not the media.

Part 2

The Power of Writing in Online Communication

CHAPTER

6

From Letters to Online Writing

Writing has had transformative powers on human cognition, society, and organizations to an extent that now we cannot imagine a world without writing. At the core of writing's power are four mechanisms—objectifying, addressing, reflecting, and specifying. They are central to the expression of emotions and development of relationships—personal, in the case of Virginia Woolf, or professional, in the case of the Hudson's Bay Company. These mechanisms are also key to understanding how scientists such as Einstein and Cartan, Descartes, and Madame du Châtelet have been able to develop knowledge with collaborators who were often in different cities or countries, and how employees and managers in complex distributed organizations such as the Hudson's Bay Company were able to coordinate though often never meeting and exchanging only one letter per year. Because they allow for the expression of emotions and the development of knowledge, these four mechanisms—objectifying, addressing, reflecting, and specifying—were essential in the development of communities such as the Republic of Letters. Through their letters, scholars such as Mersenne and Madame du Châtelet were at the center of the European intellectual and scientific life without having to travel much and, in the case of Madame du Châtelet, while officially being an outsider to the scientific community.

Yet, what do these mechanisms mean for today's communicative practices? Are they still useful in thinking about the wide array of media and technologies that are continuously evolving through the practices of billions of users? As of January 2011, there were 600 million active Facebook users (Carlson, 2011); about 200 million tweets a day sent by Twitter users, which could fill a 10-million-page book[1]; and about 1.9 billion e-mail users (in 2010), who sent 294 billion e-mails per day (in 2010).[2] On and on goes the list of statistics showing how we are "always on," interacting online, virtually in a "flat" world where distance seems to have been tamed and maybe even annihilated, a global village in which global souls (Iyer, 2001) and other neo-nomads (Abbas, 2011) stay connected with faraway others while wandering from one airport to another with their fellow international commuters and working while sitting on the beach with their families and friends.

Some scholars, reflecting on these new ways of interacting through electronic communication media such as e-mail and social media, often using mobile devices, started talking about a "second orality" (Bolter, 2001). This second orality presupposes that some electronic media (Skype, YouTube, etc.) can allow us to go back to a primary form of orality that had existed before the invention of writing. It suggests that electronic media, especially because of the audio and visual elements incorporated in many of them, offer richer and more vivid interaction patterns than the written mode and that these types of interaction can be as vivid as face-to-face encounters. This brings us back to the imagination about writing that we evoked in Chapter 1—the fantasy that we can move away from writing, which is at best only a *second* best, an imagination that goes hand with hand with the imagination that space and distance can be annihilated. If we could erase distance thanks to electronic media, we could also free ourselves from writing.

And yet, we still do write a lot. In fact, we are writing and reading more than ever (Baron, 2008; Carr, 2010; Martin, 1994), to a point that organizations are overwhelmed with documents and the number of pages printed has increased tremendously, leading some to talk about the myth of the paperless office (Sellen & Harper, 2002). Moreover, many of the new technologies—e-mail, SMS,[3] Twitter, Facebook—are mostly text based. Of course, the style and ways of writing have evolved, just as they have since the invention of writing. It remains that today we are still writing a lot, and we still enact writing mechanisms as we develop knowledge, express emotions, and build communities through electronic media.

In Part 2, we show not only that the mechanisms of writing *can* still be enacted (because they are intrinsic to the writing mode) but also how their enactment changes when people communicate by means of new media. We also examine the impact of these changes on the three powers of writing—for expressing emotions, developing knowledge, and building communities. Investigating the power of writing in terms of emotional expression, knowledge development, and community building is particularly timely. There is agreement among scholars and practitioners alike that both the ability to develop a sense of trust and shared identity within organizations (Wiesenfeld, Raghuram, & Garud, 1999) and knowledge sharing (Powell & Snellman, 2004) are crucial to the success of organizations. Informal organizations, such as online communities of various sorts—support groups, interest groups, or knowledge-producing groups such as free and open source software development—also depend on sharing knowledge and establishing trust among highly dispersed individuals.

To avoid being overwhelmed by the profusion of new media and the features they afford, we chose, in the first part of this book, to focus on letters in order to identify and examine the mechanisms at play in the process of writing. Yet, we cannot and do not want to ignore the changes introduced by the proliferation and ubiquity of new communication technologies, such as e-mail, blogs, and platforms for social networking, which are visibly influencing the nature of organizational communication and triggering various debates among organization, communication, and information systems scholars about the capabilities of various media of communication in terms of emotional expression, knowledge sharing, and allowing access to the public sphere. Indeed, the numerous studies comparing the performance of communication media have produced a plethora of contradictory findings (Byron, 2008; Constant, Sproull, & Kiesler, 1996; Daft & Lengel, 1986).

Our aim in the following chapters is to show that the mechanisms of writing we defined and presented in previous chapters can help us understand contemporary communicative practices as well as resolve contradictory findings in the literature about the powers of writing. We believe that the mechanisms we defined—because they are not dependent on a specific medium—can help us analyze emerging communicative practices with the new media, integrating their changing features into our theory, yet not being constrained by them. In this second part of the book, we examine more directly the way in which changes in technology's infrastructure and features impact writing and its mechanisms. Our aim is to look at what we lose as well as what we gain in terms of the powers of writing. To do this, we will present several case studies of technology mediated communication—emotional exchanges (called "flaming") among open source software developers; collaborative knowledge development on OpenIDEO, an open innovation platform for social innovation; and the emergence of a sense of community in a public forum on knowledge management. We also draw on a series of interviews we conducted with professionals from various fields (publishing, media, consulting, architecture) to illustrate how the mechanisms of writing are enacted with new media. Before we examine the evolution of the four mechanisms of writing in the context of new media and communication technologies, we will discuss how writing is a deeply sociomaterial practice that has constantly evolved since its invention.

THE SOCIOMATERIAL PRACTICES OF WRITING

The importance of materiality in understanding social and organizational practices has recently become a topic of study for organizational scholars—in particular, those interested in technology and work in organizations. Studies of organizational practice (Fayard & Weeks, 2007; Orr, 1996; Pentland,

1992) have shown that practice is always situated in the sociomaterial environment and remind us that we need to take into account the influence of the material environment and the artifacts used in a practice to understand it. Recent studies (particularly in the context of studying the implementation or use of technology in organizations) have highlighted how social practices are necessarily bound up with the material means—buildings, artifacts, technological devices, infrastructure—through which they are performed (Leonardi & Barley, 2008; Orlikowski & Scott, 2008; Pinch, 2008; Suchman, 2007). Similarly, we believe it is important to be aware of the sociomaterial nature of writing.

In Part 1, we focused on historical letters and, to a certain extent, bracketed the media in order to identify and investigate the mechanisms of writing and their role in the three powers of writing. Yet, this focus on writing as a mode of communication for analytical purposes does not mean that we are ignoring the importance of the materiality of the media through which writing is performed. Both matter. Indeed, writing with a stylus on a wax tablet is a radically different experience than writing with a quill on a parchment, or with a pen on a notepad, or, lately, typing on a keyboard. Depending on the media used, the experience of writing, the practices, both social and material, as well as the enactment of the writing mechanisms will differ. The French historian Henri-Jean Martin in his beautiful book *The History and Power of Writing* (1994) showed in detail how writing has gone through countless changes since the first occurrences of writing in Mesopotamia—both in the ways it was materially enacted and in the social implications of these various enactments. For example, Martin explains how a class of clerks who knew how to write emerged in France and started working for the kings in the 14th century, and how this led to the development of a bureaucratic structure, still visible today in the French society and administration. He also shows how in the 14th century the merchant class in Italy became aware of the importance of writing and realized that their children should learn more than reading prayers and singing psalms; they should learn how to read fluently, write fast, and count accurately. They therefore pressed municipal governments to develop schools for their children. As public records became more widely used, "'the logic of writing' more and more tightly imposed its sometimes Kafkaesque logic" (Martin, 1994, p. 287). Martin also described at length how the various technological inventions, such as those associated with the production of paper, influenced writing. Originally, paper was not available or was very expensive, and parchments made of animal skins were available in limited numbers. Hence, often people wrote on small pieces of parchments, which had sometimes been reused several times and which limited their ability to express fully and in detail their ideas and emotions. All the above examples show how writing is materially and socially situated and how its material enactment shaped social practices.

The medium and tools to write have evolved over the centuries, but the speed of change has radically increased over the last decade, and new media and features occur regularly. As Martin stated, "Today's specialists…are aware that writing is in constant evolution and that changes depended upon three things: the writing instrument utilized and the size of its tip; the writing material and its position in relation to the writer; and the order in which marks were made on the sheet" (Martin, 1994, p. 61). In fact, even the type of keyboard—writing on a desktop keyboard in the office, on a laptop in a coffee shop, or on the phone—makes a difference. Indeed, all the people we interviewed about their writing practices (managers and professionals in various fields such as media, technology, publishing, architecture, and consulting) insisted on how different it was to write an e-mail on a laptop versus a smart phone. All these experiences are writing experiences, but the medium used to write does matter. Obviously, the physical characteristics of the various media—size, ease of holding in hand, degree of comfort—impact the content of the written word. For example, these differences would influence the type of e-mails the interviewees would write in terms of tone, style, and content. As one interviewee noted, "I found that adopting a shorter style for mobile influenced my e-mail writing style for computer, which became briefer as well." Several interviewees stated that they could in fact recognize if an e-mail was sent from a smart phone because of its style. Reading practices are also influenced, as many do not "read" e-mails but just scan through them when checking them on their mobile devices. These material features and attending practices might lead people to stop writing long, reflective e-mails in which full trains of thought are developed, and thus could jeopardize writing's creative power.

Acknowledging that the social and the material are constitutively entangled led us to revisit the four writing mechanisms introduced in Chapter 2 in order to examine how the new material context—increased bandwidth and access, mobility of the devices, multiplicity of the platforms—influences the enactment of writing mechanisms and practices.

CHANGES IN THE ENACTMENT OF THE WRITING MECHANISMS

One of the puzzles that triggered the research we present in this book was current communicative practices and the many debates that they have fostered, and so it is important to ponder how technological changes might affect the mechanisms defined in Chapter 2 and to keep in mind that technology is constantly evolving, adding new versions, new platforms and devices, and new features, increasing the speed of connection, mobility, and so forth. We can organize these changes under two main categories. First are the changes in the infrastructure: increased bandwidth and speed, which make many of our online written communications quasisynchronous (this quasisynchronicity is also enhanced by the development of WiFi and the use of mobile devices, which allow people to be "always on"). Second are a number of technology features. For example, the ability to copy people on a message (and, to a certain extent, to forward that message) increases the number of readers and thus makes it difficult to address each one. Also hyperlinks are often included as a proof or an illustration of a point; in that sense, they provide additional knowledge. Add to these the ability to interweave replies in the original message, which supports a dialogue and allows people to respond to

NEW MEDIA AND THE BLURRING OF BOUNDARIES BETWEEN WRITTEN AND ORAL COMMUNICATION

One consequence of increased synchronicity, as well as of technical features of new media, is the increasing blurring of the boundaries between oral and written communication. The emerging communicative practices supported by new media and devices such as presentation software and smart phones have led to a more oral style of communication (Baron, 2008; Bolter, 2001). For example, messages on smart phones, while "written," tend to be interpreted as a telephone call (Mazmanian, Orlikowski, & Yates, 2006). Indeed, smart phone messages are often short, not analytical, and have little relationship management, partly because of the fast pace of exchanges, partly because of the physical constraints of the interface and the context of use (e.g., a meeting when one's attention is divided). Similar communicative practices were already highlighted by studies of e-mail and online forums (Galagher, Sproull, & Kiesler, 1998; Rice & Love, 1987; Yates & Orlikowski, 1992), but they are becoming prevalent with the use of wireless devices. In both of the previous examples, despite the written appearance of the communication (bullet points on a slide, projected or printed; words typed on a screen that can be retrieved and printed) the communicative practices tend to be more oral, in the form of short sentences devoid of analysis.

In this context of new media and new practices, the powers of writing have come into question. Statements in the press, anecdotal evidence, and our interviews all convey a general feeling that we are losing something, but it is hard to pinpoint exactly what. Part of the loss has to do with the skills associated with old-style writing, which involved drafting, reflecting, choosing the right word, and building a complex and nuanced argument. Lately, companies in diverse fields such as law or architecture have started sending their associates to writing workshops, as was pointed out by one of our interviewees, the founder and director of an architecture firm, and as quoted in a *New York Times* article: "I can't count all the lawyers who say their firms have organized remedial classes for all the associates who can't write."[4]

LETTER AND E-MAIL GENRES

Letter writing has a long history—from the great epistolary exchanges of antiquity, such as Cicero, Seneca, and Pliny the Younger, to famous personal correspondences such as Abélard and Éloise in the Middle Ages and Emily Dickinson, Virginia Woolf, or Franz Kafka in the 20th century; from the 17th century, where rules for letter writing, influenced by rhetoric and defined by an elite, emerged, and the Age of Enlightenment, where letters became the most achieved form and representation of liberty of thought (Grassi, 2005), to the development of journals, which were originally collections of scientists' letters and letter essays that served in place of complete treaties; from the letters expressing the commands on military, administrative, and political issues in the ancient Near East and Greece (Bazerman, 1988) to the development of business and administrative letters and memos in the 19th and 20th centuries (Yates, 1989).

As letters evolved, various writing practices and genres of letters emerged. Hence, whereas we often think of letters as one single genre, the 19th-century handwritten and well-written letter being a prototypical letter, there were many different types of letters—short ones, such as billets, which were very similar to today's text messages; informative and formal ones, such as the memo; very long and intricate ones which were part of complex exchanges lasting for weeks and even years; and personal ones. This variety is still found in today's e-mails, which are not always very short and to the point. In fact, a typology of e-mail—which was seen by all interviewees as a main form of writing in their work practices—emerged from the interviews. Our interviewees mentioned a range of e-mails they send and receive that map very well onto the range of letters we describe above. First are the short e-mails such as text messages that are written on smart phones and feel more like a chat (are more oral). These e-mails do not aim to create any knowledge but to ask quick questions (e.g., "Are we still on for lunch today?" "10 is perfect. Where is your office?") or provide practical information ("Be there in a few. Apologies."). Second are formal e-mails that were compared to memos. These e-mails contain briefs for projects or replies to important questions and they often come from or are addressed to clients and/or senior managers. These e-mails are written from laptops or desktops and they illustrate a certain aspect of objectifying. In such formal e-mails, for example, the client objectifies the problem to be solved and defines the scope of the project. The e-mails are often printed, and they are always kept. In a sense, they are the starting point of the knowledge-development process. Third, there are e-mail exchanges, which can last for several weeks and sometimes involve several people. They are described as "ongoing conversations" that often start as a discussion of the next steps for a project or to prepare a meeting. They often include attachments such as presentations, electronic documents, spreadsheets, and pictures or videos. Last, some interviewees also mentioned long, personal e-mail exchanges—complex and full of nuances, just like the personal letter correspondences of the old days.

specific points; the structural inclusion of background information on the sender (name, time) and the object of the message in e-mail, which provides people with a quick overview of the context; and the cut-and-paste feature in word processors or presentation software which has a significant impact on writing practices, increasing the speed of text production and collaborative writing but also presenting the danger of reproducing ideas rather than fostering critical and innovative thinking. These are but a few of the features that influence and shape our communicative practices—in particular, writing practices. In some cases, these changes lead to a blurring of the boundaries between oral and written communication, as our written messages "sound" more oral (see the sidebar New Media and the Blurring of Boundaries Between Written and Oral Communication).

However, it is important to recognize that, despite the changes, the four mechanisms of writing can still be enacted in various forms in electronic documents and presentations, e-mails, and blogs. In our view, successful examples of emotional expression, knowledge development, and community building with today's media are largely explained by the enactment of writing's mechanisms. While new media might afford new features and practices, we always have the freedom to use the technology in different ways. Indeed, people tend to write shorter e-mails with little or no relationship management; and they tend to reply immediately to e-mails without taking the time to think about it; and yes, there is a propensity to use cut and paste while working on a document. However, there is no determinism inscribed in the technologies, and the accompanying practices afforded by the features of the media are flexible. For example, one can take the time to draft an e-mail and send it a couple of hours or even a day later. One can also decide to start a new document from scratch. Moreover, as we will see in the following chapters, some of the features of new media can facilitate the enactment of the writing mechanisms. Finally, beyond the variations introduced by the new media, there are also a lot of similarities, not only in the enactment of the writing mechanisms but also in the range of genres that one can find in letter writing and in e-mail writing (see the sidebar Letter and E-mail Genres).

In the following section, we will revisit the four mechanisms of writing introduced in Chapter 2 and show how their enactment is supported, or hindered, by new media.

Objectifying

It is through the act of *writing it down*—which often involves scribbling and drafting—that we are able to *objectify,* to articulate an idea or an emotion. The process of writing helps clarify one's ideas and feelings so that what was initially amorphous is articulated and takes on a life of its own. Through objectification, an idea is expressed more clearly and can be reflected upon, specified, and modified by the writer as well as by others with whom he or she can share the idea. The objectifying process is still afforded by various new media. Yet, because messages tend to be shorter, and a "chat" style (more like a quick, oral, back-and-forth conversation) has emerged, in many cases the process of writing down tends to be degraded. Objectifying tends to become almost synonymous with the output—that is, the trace that is left—with an ability to easily archive all messages (through increased space but also because of sites like Dropbox and Google) and a facilitated ability to search these archives. Hence, it seems that the process element of objectifying is diminished.

At the same time, the dialogic process of writing can be accomplished more easily than in handwritten letters because of the ease of interweaving in the text of e-mails and the commenting features of blogs and other online platforms. Thus, new media do not prevent us from enacting objectifying and, in some cases, such as commenting and replying, can even support the objectifying mechanism. In fact, all our interviewees insisted on the importance of "writing down"—on paper or on a computer (desktop or laptop, yet, very rarely, on a hand-held device)—as a way to articulate their thoughts or emotions. Several mentioned that they used the draft feature of an e-mail not only for e-mails they sent but sometimes to write down their ideas. They either share their ideas with others by sending an e-mail or document, or they just go back to their ideas to reflect on them and develop them further. Several interviewees talked about the importance of the writing process in clarifying ideas and how it could even take them to a place not known at the beginning of the articulation process. These examples highlight the role of objectification in enacting the creative power of writing and show that this power can still be performed in online contexts.

Addressing the Reader

When we write a letter, speech, or book, we always write *to someone,* trying to adapt our text to the (potential) reader's perceived needs, knowledge level, attitudes, and interests. Addressing has become crucial both for expressing emotions, for developing knowledge, and for building communities. However, addressing is sometimes jeopardized by the near synchronicity brought about by increases in bandwidth and speed. For example, in distributed teams, because members can receive a reply in a matter of seconds from teammates located thousands of kilometers away, in other countries, sometimes on other continents, they tend to forget about cultural specificities, context, and time-zone differences. Their feeling of perceived proximity (Wilson, O'Leary, Metiu, & Jett, 2008) makes them forget to adapt their e-mails to their teammates' context, knowledge, and cultural perceptions. This often creates tensions because contexts vary and cultural differences still exist. Furthermore, differences in context (in terms of cultural norms, incentive systems, political agendas, or time-zone differences) tend to be overlooked when one writes to not only one person but to several (sometimes 10 or more), all located in different regions. Even when one is aware of such differences, it is challenging to address all these correspondents properly.

At the same time, new media in some ways support addressing by providing some contextual elements, such as the name of the recipient, the time the message was sent or posted, and the object of the message. New media also support addressing by providing features that are not written, such as photos, links to websites, videos, mood messages (little messages on instant messaging platforms, where we can tell our friends and colleagues our mood or leave a quote or comment for everyone to see). Moreover, one of our informants explained that she often drafted her ideas by writing them in an e-mail saved as a draft. She noted that it made a difference to write her ideas as an e-mail addressed to someone rather than just as a note to herself, as it forced her to be more specific. Once again, we see that new media do not prevent us from enacting the addressing mechanism and in some cases they even afford some new ways of enacting it.

Reflecting

Because ideas and emotions can be objectified through writing, the writer as well as the reader can read the text (for the writer, it can be in multiple drafts) several times and reflect. Our interviewees talked about how "writing frees reflectivity." They insisted upon the importance of writing for generating as well as developing new ideas; they described writing as a clarifying process in which they extensively used the drafting feature that is made easy in e-mail systems and also in documents and presentations.

At the same time, several characteristics of the new media are less conducive to the enactment of the reflecting mechanism. Thus, the increasing speed of communication—the faster reply cycle (as well as, for many of us, a feeling of being expected to reply quickly), the size of the screen when using hand-held devices, and the context (often in busy places, on the move) can lead to fewer reflecting opportunities. Moreover, the frequent interruptions we face in the form of new e-mails, hyperlinks, and tweets further stunt the reflecting process. However, the lack of reflectivity is not intrinsic to new media but is related to the nature of the Web. As Nicholas Carr showed in this book *The Shallows,* the Web has become a "universal medium" that plunged us into an "ecosystem of interruption technologies" (Carr, 2010). Thus, even though we are oftentimes connected to the Internet when writing an e-mail, an electronic document, or a presentation and thus potentially interrupted, this is not intrinsic to the new media but to the Internet. We can write documents and presentations (and even e-mails, in some cases) while being off-line.

Our interviewees were keenly aware of the dangers posed to the enactment of the reflecting mechanism. Furthermore, they have developed strategies to circumvent this potential loss. For example, they used their smart phones to check e-mails but mostly to screen them, so that when in the office they could focus on the "more complex ones, those which require thinking." They usually put aside until a later, quieter time, long e-mails that they knew required time and attention and those that contained extra documents (either attached to the e-mail or accessible only on their computers). They tended to send from their phone only short e-mails that did not require reflection and the articulation of ideas. These strategies suggest that as people feel the limitations of new media, they also are able to carve for themselves situations when reflecting can be enacted.

Specifying

As we try to articulate our emotions and ideas, we usually attempt to be as specific as possible, adding details and using precise language. Specifying is what allows people to articulate emotions or thoughts in detail. Specifying is both strengthened and weakened in the new media. As noted above, increased bandwidth and mobile access have increased the speed of communication and have led people to write short messages, sometimes with a "yes" or a "no" or with a link to a website as a reply to a question. However, many of our interviewees noted that for them this kind of e-mail, often referred to as SMS (text message) or chat type, is not perceived as written. This means that they may not search for the precise nuanced word that would capture their ideas, and that they may not give the needed details. Oftentimes, the platform (a laptop or a desktop rather than a smart phone) and the context (at home or in the office rather than at the airport or during a meeting) are elements that have an influence on whether the specifying mechanism is enacted or not.

On the other hand, the ability to comment inside the body of the message allows us to reply specifically to a particular point. Moreover, the ability to cut and paste and to insert links and images strengthens the specifying mechanism. Interweaving practices also help in making specific replies to particular points and questions raised in the e-mail or the post. As is the case for the other three writing mechanisms, we see that specifying can still be enacted through writing, at least with some technologies. Hence, while new media and communication technologies provide new material and social contexts of use that influence communication practices, they do not make the enactment of writing mechanisms impossible.

In Part 2 of the book we analyze the enactment of the three powers of writing—expressing emotions, developing knowledge, and building communities—in contemporary communicative practices. We analyze what has been gained and lost by the advent of new communication technologies through case studies and interviews. Naturally, the structure of Part 2 reflects the structure of Part 1. Thus, Chapter 7 analyzes how writing-based modern technologies support the expression of emotions, in particular in an open source software development community. In Chapter 8 we discuss how the mechanisms of writing can explain the development of knowledge in computer-supported collaborative work, such as the OpenIDEO open innovation platform. Chapter 9 shows how communication technology and online platforms enable the development of online communities as well as the participation of outsiders such as women in open source software development projects. In the last chapter, we discuss the implications of our research—the creative dimension of the writing process—for today's organizations, where creativity and constant innovation are continuously sought after and suggest new directions for future research.

CHAPTER

7

Expressing Emotions and Developing Trust Online

The Internet may be virtual, but the emotions people express and experience while riding its virtual waves are quite real, sometimes very painful and full of rage, sometimes joyous and enthusiastic, sometimes teeming with raw energy and desires. It is a passionate place, just like the old-world of letter exchanges we examined in Part 1. The virtual realm is the scene of making friends, of falling in love and having affairs, of flourishing affinity as well as of hate groups. It is a safe place where cancer patients find solace expressing their most intimate feelings and fears and are responded to with kindness and comprehension. It is a place where children who were adopted and their parents find like-minded individuals who are attuned to their most cherished or difficult emotions—a place where members of social movements, such as advocacy networks organizing slum dwellers to obtain housing (Sassen, 2002), can express their beliefs in and support for various causes, find allies all around the world, and thus increase the reach of their communities.

Most of the time, these emotions are expressed through writing. While images (photos, videos) that are easily inserted into electronic communication play a role in the spurring of emotions conveyed through electronic media, they are rarely used. The writing mode seems powerful enough for emotional expression: As Sherry Turkle (1995) has shown in her book *Life on the Screen*, people who interact only through writing in multi-user dungeons (MUDs) often describe the emotions they feel in their online interactions as more "real" than the emotions in their off-line life.

Given our analyses in Part 1, this is not surprising. As illustrated in Chapter 3, the mechanisms of writing play an important role in the expression of emotions. We saw that emotions could be expressed in writing in both organizational and interpersonal settings, with Hudson's Bay Company (HBC) employees expressing frustration and criticism while attempting to build accountability and trust, and with Virginia Woolf's correspondence displaying a wide array of subtle and nuanced emotions.

But do contemporary writing practices and media show similar emotional displays? What has changed and what, if anything, stays the same? So far, empirical research on the capabilities of new media and online communication to support the expression of emotions and the development of relationships has produced a host of contradictory findings. In this chapter, we show how conceptualizing online communication—e-mail, blogs, forums—as instances of written communication can help us move beyond the contradictory findings of empirical research and help us understand the constant evolution of contemporary communicative practices. We illustrate online communication's emotional power with a case study of "flaming" messages (hostile and insulting interactions between Internet users) in an online community of Linux developers.

A RAGING DEBATE: EMOTIONS AND ONLINE INTERACTIONS

It is probably fair to state that so far we lack a reliable understanding of emotional expression in online interactions. Scholars adopt theoretical stances that focus on various aspects—media features, interactants' characteristics, context—of the phenomenon, and then, not surprisingly, they often produce contradictory results. A main stream of research, media richness theory—which we reviewed in Part 1—focuses on the characteristics of the media and holds that because of their lack of nonverbal cues and their asynchronous nature, media such as e-mail are poor at conveying rich and nuanced emotions (Daft & Lengel, 1984, 1986). Assuming that only face-to-face communication can provide a wide array of cues and language usages that enable exchange partners to express themselves fully and form a sense of one another, this stream of research concludes that emotions cannot be expressed in written-based media such as letters or e-mail (and by extension, blogs and forums).

While media richness theory implicitly refers to emotions and relationships, the main theory that has addressed this aspect of communication is social presence theory (Short, Williams, & Christie, 1976). Similar to media richness theorists, the proponents of social presence theory criticize computer-mediated communication for its poverty, arguing that a medium's social effects are principally caused by the degree of social presence it affords users. *Social presence* refers to the awareness that the communicator has of an interaction partner. The degree of social presence in an interaction is determined by the communication medium, which is characterized and ranked (implicitly) along the same continuum as the one suggested by media richness theory. Therefore, an e-mail conveys less social presence than a face-to-face interaction (which provides the maximum social presence) because it misses nonverbal cues such as facial expressions, tone of voice, gestures, direction of gaze, and posture. Again, the assumption is that writing cannot convey one's state of mind, context, emotional and ideational state, or one's questions, doubts, tribulations, hypotheses, conjectures, and hopes.

Furthermore, the presence of the interaction partner is always considered as beneficial for the expression of emotions. The possibility of the other's presence being distracting is not considered, nor is the possibility that in the other's absence one can better try to sort out complex emotions, and appropriate and understand them. Consequently, writing is assumed to be an impoverished mode of communication, which cannot support well the expression of emotions. These theories, therefore, indicate that because computer-mediated communication is task-oriented and impersonal, it does not support emotional expression and relationship development. However, as we show below, the empirical evidence for these claims is mixed in two aspects.

First, scholars in these traditions make a strong distinction between positive and negative emotions. Thus, while positive emotions seem to conform to the theory, negative emotions behave in a radically different way in the sense that they seem to be amplified in the online setting. For example, because computer-mediated communication eliminates cues that individuals usually use in face-to-face situations to convey trust, warmth, and other types of interpersonal emotions, negative emotions get amplified in written media such as forums and e-mail and therefore misunderstandings are more frequent in these media (Sproull & Kiesler, 1991). This phenomenon is also amplified by people's tendency to interpret even neutral e-mail messages more negatively (Byron, 2008). Thus, not surprisingly, Sproull and Kiesler (1986) argued that when social context cues—including aspects of physical environment and nonverbal behaviors—are missing, more excited and uninhibited communication (such as flaming—i.e., hostile and insulting interactions via e-mails or in online forums) can ensue. In a similar vein, Kraut, Steinfield, Chan, Butler, and Hoag (1999) and Putnam (1995) argued that e-mail and online forums do not support the creation and development of emotional bonds among people. Second, the empirical evidence is mixed, even with respect to terms of positive emotions. Thus, in contrast to the view that electronic (written) communication is dehumanizing and less rich in social cues, a set of studies has shown that computer-mediated communication can support emotional communication (Johansen, DeGrasse, & Wilson, 1978; Rice & Love, 1987) and that people may in fact engage in more intimate exchanges online than in face-to-face interactions (Tidwell & Walther, 2002). On the basis of such studies, Walther (1996) has argued that mediated interactions can be as rich as face-to-face interactions. He developed the social information processing theory, which claims that the limited bandwidth offered by computer-mediated communication (compared to face-to-face) can be overcome as people get acquainted to the medium and become familiar with their communication partners (Walther, 1995, 1996; Walther & Burgoon, 1992). According to this theory, computer-mediated communication does not differ from face-to-face communication in terms of the capability of social information exchange but rather in terms of a slower rate of transfer. Thus, Walther (1995) reported that computer-mediated groups become less formal and less task-oriented over time. Furthermore, researchers have also found that online networks of electronic support groups offer intense emotional and practical support (Galagher, Sproull, & Kiesler, 1998).

Similar contradictory findings were produced by studies of trust-based work collaborations among faraway partners. While numerous studies have highlighted the difficulty of building trust in virtual teams and distributed organizations (Nardi & Whittaker, 2002; Olson, Teasley, Covi, & Olson, 2002; Wiesenfeld, Raghuram, & Garud, 1999), other empirical studies showed that trust in computer-mediated teams can reach levels comparable to those in face-to-face teams over time (Jarvenpaa & Leidner, 1999; Wilson, Strauss, & McEvily, 2006).

In our view, the social information processing theory gets closer to explaining otherwise puzzling phenomena such as Internet friendships and romances, or complex work collaborations involving trust. Still, the possibility that positive emotions get amplified is not explored; as if sharing deep emotional experiences and developing strong bonds—work or nonwork related—with people one has never seen and will never meet, does not exist. At the same time, we argue that an important factor in explaining such phenomena is that online communication is just an occurrence of written communication. In our view, the writing mechanisms we defined can more richly explain the conflicting findings about emotions and trust in the online context than can merely referring to intrinsic limitations of the medium, such as the lack of social cues or by adding social dimensions to the analysis. In particular, we argue that the changes in the enactment of the writing mechanisms may explain, in part, the specific ways in which emotions get expressed and trust gets built in online contexts. In the next section, we will illustrate these claims with a case study of flaming in a Linux development forum, showing how developers were able to express both negative and positive emotions by enacting the mechanisms of writing.

FLAMING AND EXPRESSING EMOTIONS IN A LINUX FORUM

We chose a Linux forum for several reasons. First, keeping with our focus on work-based collaborations, we chose a productive setting devoted to developing software products (as opposed to an online support group such as a cancer support group). Second, it is a context that is at the forefront of distributed work collaborations because most interactions take place online. Third, the free/open source software movement contributes to the shaping of the future both in terms of technical products and in terms of collaborative practices (O'Mahony & Ferraro, 2007); this should enable us to identify the changes in the enactment of the mechanisms in this new technological context. (For a short description

FREE AND OPEN SOURCE SOFTWARE DEVELOPMENT COMMUNITIES

The free and open source software communities collaboratively develop software that allows users to use it free of charge, change it, and redistribute it. Communities of developers spanning the earth and communicating largely via online forums have developed successful products, such as the Linux operating system, the Apache Web server, and the PERL language (for more details on the organization of free/open source communities, see Kogut & Metiu, 2001). Thus, thousands of developers scattered throughout the world have developed and are maintaining Linux, which is currently used by millions of people in all countries. Just like other production-oriented organizations, the free/open source software community is mostly concerned with producing results; while HBC was focused on trade, open source aims to produce stable, robust software.

The free/open source software community is an informal organization, a transnational community rooted not only in shared interests or collective projects but also on shared values (Djelic & Quack, 2010). While its main goal is clearly devoted to knowledge sharing and development, it is also a setting of strong emotions and emotional display. Indeed, these developers care deeply about the quality of their work as well as about the values of free software. The philosophy of the movement is to give freedom to computer users by replacing proprietary software with free software. The ideology of free/open source software comes in two flavors, reflected in the combined name of the movement. Regardless of these two flavors, both stand in marked contrast with the prevailing business emphasis on software patents and secrecy of proprietary software.

One of the consequences of developers' passion for their work is the emotional tone of many messages. Sometimes the strong emotions lead to what has been called "flaming—that is, messages that speak incessantly and/or rabidly on some relatively uninteresting subject or with a patently ridiculous attitude (*The Original Hacker's Dictionary,* posted at http://www.dourish.com/goodies/jargon.html).

of the purpose and functioning of free and open source software development, see the sidebar Free and Open Source Software Development Communities.)

The Linux Forum

The forum we studied consists of Linux kernel members (the kernel is the operating system's main component, the one that ensures the connection between the data processing in the hardware and the applications). This large group—there are hundreds of developers from around the world who contribute code to the kernel—know and respect one another's work. Most of the members have industrial or academic backgrounds and are well versed in their respective fields. They are scattered around the world, and in our data they refer to meeting face to face (e.g., at a conference) only a few times.

Linux developers in this forum do not refer to themselves as belonging to a special community corresponding to the forum; they see themselves as part of the larger community of the "Linux developers." References are also made to other communities such as the Solaris or Java developers. Contributors post pieces of code they developed and provide highly technical arguments and comments about why or why not a particular piece of code would work. Most of the time, the members respect each other's opinion and are glad to see things from a different point of view. They do not hesitate to ask for help from people they consider experts. Some members get defensive when a person they support is attacked. Through the discussion, it is obvious that this is a collection of experts who possess strong feelings about their beliefs.

While in this forum members enact all three powers of writing—emotions, knowledge, and community building—we focus in this chapter on emotions. In fact, at several points the discussion rages, and flaming messages emerge between members debating about some technical solutions. Though many members try to dissuade them from using rude language, anger is usually expressed through swearing. However, several members take on the role of a forum administrator, reminding the errant poster that it is a public forum that should be respected and that the personal comments must be moderated. The flaming discussions usually end without being resolved and with posters being distracted by involvement in multiple topics simultaneously. Thus, this forum is the site of intense emotions expressed in strong language. Their emotions are related to the developers' knowledge, as they try to convince others that the solution they propose is best. Such debates are not unlike the argumentative exchanges that sometimes arise in scientific communities (the recent clashes over climate change is one of the latest examples).

The Flame

We did an in-depth analysis of a specific thread called "devfs-why not?" started by R. K.[1] on the Linux kernel forum on Tuesday, April 11, 2000. It consists of 124 messages and lasts for 4 days. Twenty-five members were involved in this thread, with 10 being very active. The debate was induced by the inclusion of devfs (devfs is an abbreviation for *device file system*) as a replacement of /dev. UNIX or its clone implementations provide (traditionally) their users with an abstract device path called / dev. Inside this path, the user finds device nodes, special files that represent devices inside their system. The device nodes enable the developers to access the devices as files and hence did not require using any special APIs.[2] The devfs is a Linux solution to many problems that the traditional /dev mechanism has. This led to clashes between two opposing camps. The posters seemed to be mostly employed in academia or in industry in Canada and the United States (as indicated by their e-mails).

Though the initial message is a request for more information, the thread soon becomes a "/dev vs. devfs" discussion. With messages flying back and forth, the discussion quickly escalates into a flame war between the two groups. While one group claims that devfs can aid in solving multiple problems, the other despises its "ugliness" or lack of finesse. Posters make a lot of personal insults, including attacks on their opponents' lack of professionalism, their limited knowledge, their low IQ, and even include downright name-calling. The original topic is soon forgotten, and the discussion branches out into multiple issues. Finally, it ends without being completely resolved, despite some differences being settled over the course of the four days.

Our analysis reveals that the exchanges in this flame war allow the developers to freely express their opinions and beliefs, to advance their knowledge, and to ultimately develop a better product. Our analysis also shows how writing allows developers to express both negative and positive emotions through several practices: criticizing the work and views of others, appealing to rationality, and engaging in a true dialogue.

Expressing Negative Emotions: Criticizing the Work and Opinions of Others

The prevalent emotion in the flame is anger, mostly arisen from the frustration of not having one's viewpoint accepted by others. Anger and frustration are evident in the use of insults, of all caps, of sarcastic humor, and of emoticons. Here is how one developer reacts to the change to devfs: "Why are you then building all this crap on top of the parallel tree in the FIRST PLACE?!" The developer expresses his puzzlement at the introduction of the devfs, puzzlement due to his fundamental criticism concerning this change and the rationale for it. Clearly, he does not see the need for such a change. Yet, as his comment is quite general and does not specify in detail his objections, it has the potential to spark a flame, which it did. The developer writes as if making an oral comment. Yet, while an oral comment might hurt someone feelings, it can be explained and discussed quickly, thus avoiding the escalation of negative emotions. On the contrary, when the criticism is written down, it can be misinterpreted with no possibility of being clarified. In this case, as the developer writes down his criticism, he objectifies it, leaving it on the forum to be read by many yet without specifying the rationale for his criticism.

The devfs author is clearly affected by this post, because he answers immediately, expressing his anger at the attack: "JUST ONE OF THE MANY THINGS YOU'VE GOTTEN WRONG. I CAME UP WITH devfs."

Another developer has set out to fix the damage he perceives the change to devfs will inflict on the operating system, and he writes to all who have questioned his undertaking: "Until it will be finished inquiring minds are welcome to sod off." The tone of these statements is very aggressive and insulting.

In contrast with the attacks above, some of the criticism is detailed and specific while still maintaining the angry aggressive tone of a flame. Being specific in a forum devoted to knowledge development is quite expected, and often participants explain in detail proposed functions and specific pieces of code (often attached to their posts—a great example of objectifying). Here is how a developer, A.V., refers to a list of specific things he works on in order to maintain the robustness of the operating system after the introduction of devfs:

> no. it's spelled "list of the things that must be written before R.'s code will cease being a lovely way to crash the kernel". WHY DO YOU THINK I'M DOING THAT KIND OF REWRITE DURING THE FREEZE? one more time: r's support for multiple mounts is fundamentally broken. By design. It got into the tree. Fixing it requires changes to the WHOLE FSCKING LOOKUP/MOUNT CODE. YES, IT'S THE RIGHT THING TO DO, BUT IT became necessary to do _now_ (instead of doing it more or less at leisure during the 2.4 with merge in 2.5.early) precisely because of devfs. That's WHAT I'M DOING. SINCE SEVERAL WEEKS AGO. FIXING THE DEVFS-PROPER REQUIRES all this stuff done. as i've told: after (list of changes) will be done.Learn to read. And fuck off, already.

We see several of writing's mechanisms at work in this example: references to others posts and to posted lists (objectifying), reminders of the author's identity as a main contributor to the kernel (addressing), and being precise about an alternative time when the changes could be made (specifying). This being a flame though, even the technical details about the work done are interwoven with strong language and insults. Knowledge development is not a detached undertaking; it involves people who hold strong views and who are ready to defend them, including by means of expressing strong negative emotions and insults.

Appealing to Rationality: Toning Down the Negative Emotions

Later in the flame, other developers enter the conversation with a less emotional tone and try to get the focus back on the technical issues being debated. The *objectifying* mechanism is central to the expression of emotions as well as the understanding of the issue at stake. For example, the participant below refers to the fact that he can go to the webpage of one of the main actors of the controversy as well as the various posts in order to "make a judgment" for himself; he is in fact asking others to provide him with references to relevant threads:

> I have read R.G.'s webpage on devfs and it looks like a brilliant idea to me. Now, I understand that R. had to fight A LONG STRUGGLE GETTING DEVFS INTO THE KERNEL AND EVEN NOW THAT IT finally goes in, it does so in an (imho) somewhat crippled form (mounting it under /devfs instead of /dev) and there still seem to BE MANY PEOPLE OPPOSED TO DEVFS. so, i know the pro side of devfs, but i would like to read up on the objections raised so I can make a judgement for myself.i guess this has been widely discussed on this list but that must have been before I joined, so I would be glad if someone could point ME TO SOME ARCHIVE THAT HAS THE RELEVANT THREAD(S).

It is relevant to note the issues the author chooses to put in all caps, a type of font that is interpreted as shouting. The first is an acknowledgment of the flame with both sides ("A LONG STRUGGLE GETTING DEVFS INTO THE KERNEL"; "MANY PEOPLE OPPOSED TO DEVFS"). The second is a request for more traces of the various exchanges ("SOME ARCHIVE THAT HAS THE RELEVANT THREAD[S]") so that he can reflect on the arguments on each side. The trace (objectifying) helps the reflecting process: "i would like to read up on the objections raised so I can make a judgement for myself."

Several other developers appeal to their coworkers' rationality by asking them to focus on the technical aspects of the debate instead of fueling the ad hominem attacks. For example, this participant tries to reroute the discussion by reminding them the original question: "i'm not even talking about that anyway, and the discussion didn't even start that way. Can we get back to the original topic?" This other developer is also trying to redirect the discussion, avoiding emotions as well as supporting knowledge creation: "i have the feeling i'm not making my point clearly enough here since you're still fixated on filesystems. Did you even read the rest of my EMAIL?" In this excerpt, the developer is reflecting on his ability to express his point ("i have the feeling i'm not making my point clearly enough") based on his analysis of the other's post ("since you're still fixated on file systems"). He is suggesting to his interlocutor to take advantage of the objectifying and reflecting mechanisms of writing by suggesting that he should read the developer's whole e-mail.

We also see half-hearted attempts at cooling down the flame, such as in the example below in which D. P. would like A. V. to stop insulting those who disagree with him. However, D. P. does this using an inflammatory tone as well:

> Then why bother beating R. up about his code every time you CAN FIND AN OPPORTUNITY? I WOULD BE SURPRISED IF THERE WAS ANY lkml reader who didn't know that you think that the current devfs is worse than Hitler, Stalin, Pol Pot, fluoridated water, the US CENSUS, THE NEW WORLD ORDER, CANADA, AND WWW.MICROSOFT.COM, AND thus you're not really enlightening the masses as much as engaging in a prolonged and rapidly getting boring hissy fit.

This post is interesting because while it claims the developer is frustrated by the proportions of the flame, the use of sarcasm and the accusations that A. V. acts inappropriately are not likely to cool down the flame.

There are many reasons why it is difficult to cool down the flame. One is the trace left by all these messages that can be read and reread and that can make developers angry again. The other is the audience effect, the posturing involved when the poster knows that many developers are reading the posts. Another reason is the speed of exchanges; they are so quick to write and post and display that they don't leave posters room to cool off their emotions.

Engaging in a True Dialogue: The Passionate Community

The previous excerpts already give a good sense of the liveliness of the debates in this setting, of their spontaneous, uncensored character. As we explained, the participants care passionately about their work, about producing reliable, high-quality software, and about the virtues of non-proprietary software development. They are involved in frequent discussions and debates of (mostly) technical issues, such that the dialogue among them is continuous. The true dialogue in the forum is enabled not only by the hacker culture, but also by some of the technical features of the online medium.

Thus, the ability to refer to previous threads and to intertwine replies within previous messages, creating a lively dialogue, is illustrated in this excerpt:

> > LINUX CAN DO VERY BAD THINGS TO THE SYSTEM:-) Oh God…The mind boggles at the thought.:-) Giggle…

Another factor that contributes to the lively dialogue is that these exchanges are not anonymous. Kernel developers know one another because they've been working together for some time, they are familiar with one another's expertise, and sometimes also met at conferences. They all sign with their real names, so often the messages are directed to one particular person, as we saw in the exchanges above, or in examples such as "You completely miss the point of this discussion," or "SHEESH, I DIDN'T THINK I HAD TO EXPLAIN THIS TO YOU." As these examples show, despite the public nature of the forum, participants are still engaged in a dialogue with specific participants.

At the same time, the entire community is concerned about the change to devfs and thus by the ensuing flame, posters are keenly aware of the audience—the developers who do not necessarily post but follow the exchanges closely. This is obvious in the following example, which shows how in the same post, the developer addresses directly both his opponent and the entire community:

> I MAY NOT BE THE ONLY ONE WHO THINKS THAT you're making yourself look like an idiot tilting at the thousands of windmills that offend you.

The reference to Don Quixote is intended to mark the opponent as a loner with respect to the rest of the community, someone on an unattainable, even unreasonable quest.

Another element that contributes to the feeling of a true dialogue and that certainly contributed to the escalation of the conflict and to the sparking of insults is the speed with which these exchanges take place online. Indeed, the exchanges were very intense over a short period of time (four days), with little or no censorship, with people writing exactly what's on their mind, without much regard for the feelings of others.

Flaming E-mails and Emotions

Our example of flaming in a community of open source developers shows that informal virtual organizations are "emotional arenas" as much as formal organizations (Fineman, 1993). Strong, agonistic exchanges among members of knowledge communities are common. While the correspondence between Élie Cartan and Albert Einstein, which we presented in Chapter 4, was very polite and positive, in his correspondences with other correspondents, Einstein sometimes expressed strong opinions and disagreements. Thus, Einstein's letters to the physicist Max Born show that the two had many points of discord. At one point, Einstein decided to stop even replying to Born because of their disagreement about the nature of quanta.[3] What makes the difference with the example analyzed above is that in open source software development, the audience is bigger and what is written stays for all to be seen and read multiple times.

Far from being devoid of emotions, the virtual realm is full of them. Because the virtual exchanges are written exchanges, the mechanisms of writing are at work in online forums just as they were in the classical world of handwritten letters. In Chapter 3 we saw that HBC commanders had to find convoluted ways to express their frustration and criticism of headquarters in a context of strong hierarchy and traditional organizational culture. In contrast, the open source developers are peers steeped in a computing culture that is strongly agonistic: the flame, as violent as it may seem, is an integral part of the argumentative culture that is geared toward producing code of the highest quality. In this community in which flames flare up regularly, the devfs flame controversy is considered by free and open source software developers as one of the main flames in this community (hence its "war" status). Thus, in our view, the explanation for the strong emotions expressed in this forum is less technical and more social: it is the particular subculture of hacking and computing that is responsible for the high incidence of flaming in online contexts (Dubrovsky, Kiesler, Sproull, & Zubrow, 1986; Lea, O'Shea, Fung, & Spears, 1992). Specifically, the hacker subculture supposes a challenging attitude and distaste for authority and, thus, a penchant for expressing strong opinions freely, with little regard for social conventions.

This view is in contrast with streams of research that have focused on the characteristics of the interaction medium, as stated by the proponents of the reduced social cues theory (Siegel, Dubrovsky, Kiesler, & McGuire, 1986; Sproull & Kiesler, 1986, 1991). Our view is also that these two reasons—hacker subculture and media features such as increased speed—are not mutually exclusive. At the same time, we need to reemphasize that negative emotions and strong disagreements are not the only type of interactions and emotions in these communities. Strong positive emotions of belongingness to the community, of beliefs in its ideals, are also frequent; thus, one study reports that out of the 1,540 developers surveyed, only 14% said the ideology was not important (David, Waterman, & Arora, 2003), thus demonstrating the developers' commitment to their community. In the next section, we will expand our investigation of the role of writing in expressing emotions with new media. Based on interviews we conducted with various professionals, we will examine how each mechanism of writing can be enacted in online contexts and how each mechanism supports the expression of emotions.

EXPRESSING EMOTIONS AND DEVELOPING TRUST IN WRITING IN ONLINE CONTEXTS

While all four mechanisms of writing play a role in the emotional expression shown in the flame in the Linux case, flaming also emerged due to the influence of online media's affordances on the enactment of writing mechanisms. For example, the objectifying mechanism tended to amplify the flame because of the publicity of the forum and the fact that messages were "there" to be seen by all. Objectifying was thus reduced mostly to its output. Mechanisms such as reflecting or specifying were not always performed by the Linux developers. Overall, in this knowledge community, writing supported both rational arguments as well as a passionate dialogue among the members of the community. Our interviews and analysis of personal experiences, discussed below, confirm our claim that all writing mechanisms can be enacted in online contexts, although with some variations due to the different features and associated practices afforded by new media.

How the Four Mechanisms of Writing Enable the Expression of Emotions and the Development of Trust With New Media

Objectifying

The objectifying mechanism is key in the expression of emotions. Through the writing process, fleeting and immaterial feelings become objects that can be expressed, reflected upon, and shared. Furthermore, the articulation process allows the writer to clarify his or her emotions.

All text-based media—either letters or online forums—enable the objectifying process through which we can articulate emotions—for ourselves and for others. Precisely because there are few bodily cues about the recipient(s) or the writers, the written word can be very revealing of one's emotions because, in the absence of the others, people may dare express things that they would not otherwise and because the writing process helps the writer clarify the emotions for him- or herself. Of course, as we saw in the flame above, this can have negative consequences for the tone of the ensuing exchange when the emotions are frustration and anger. Furthermore, the output—it's written out there—also makes the negative emotions more "formal" and "stronger," especially as the number of people that can read the message increases (the "copying" effect in e-mails and "public aspect" of forums).

The impact of the written word on one's emotional state and the way in which readers infuse the written word with meaning is as strong as in private, one-on-one exchanges. A writer we interviewed talked about her tense relationship with the editor of her book, who was located in a different country and with whom she worked for 18 months. She said:

> Because every e-mail from [her] as the editor of this 1,200 page book needed an immediate response, I developed amazing anxiety, my computer was pinning every moment; as soon as I answered, she wrote back. Her mind being very mathematical, she was doing the grids and the design and I was doing the narrative and the text. I never got back to enjoying the communication since that episode, even though the book was a success.... None of the 9,000 e-mails was warm; they were all about detail.... I understand the effectiveness, but the coldness affected me deeply, very emotionally.

The words of this writer capture the strong impact of negative emotions that e-mails can have on one's emotional state. In a way, the e-mails became the symbol of the relationship, just like Virginia Woolf talked about some of the letters she received. Such e-mails can be kept—saved in special folders or even printed. It is interesting to note that some people seem to have transferred some of their emotional attachment to the device itself, their mobile phone, and the constant connectivity it offers them (Vincent, 2006). One of our interviewees described the sensuality and materiality of the writing experience on her mobile phone:

> The thing I do like about my phone is I can do the "wish you were here" window, no need to go to the computer. The computer doesn't have the intimacy of the phone, of holding it in your hand; the phone has a delicacy; and I use both thumbs, and I'm quite fast, and it has a rhythm that is very sexy; and the click that you don't have on a computer (now you have it on iPad but not on the computer) is very sexy too; like a heartbeat.

At the same time, the strong emotionality we saw in the flaming example and in the quotes above should not be extrapolated to other organizational settings. Our interviewees seemed keenly aware of the potential for escalation, misunderstandings, and audience effects in electronic communication in work contexts. They talked about their efforts to balance the pressure for fast responses with the need to prune their writing of emotional expression. Many insisted on being careful about using humor in fear of being misunderstood and hurting people's feelings. Several interviewees noted that when they were frustrated or angry, they tried not to reply immediately—that is, using the asynchronicity offered by writing—in order to avoid an escalation of e-mails, the organizational version of flaming. They often would draft a reply, and leave it aside, going back to it a few hours later or even the next day. They would often modify the e-mail one more time before sending it. In some cases, they would even not send it at all. The simple fact of writing down their emotions and objectifying them helped them calm down and avoid direct confrontation.

Addressing

As we showed in Chapter 3, the writer's understanding of complex emotions is facilitated by writing them *to someone*, by addressing the reader. Similarly to letters, online interactions provide a space where participants can engage in a dialogue. Hence, most of the posts in electronic groups are "conversational" (Galagher et al., 1998). The online context facilitates the enactment of the addressing mechanism because the posts indicate the sender's name, the time, date, and subject and often provide a link to the sender's profile. Even in conditions of anonymity though, online interactions still are directed to a distant other, who one imagines and whose subjectivity one tries to reach. Galagher et al. (1998) reported that because online interactions offer some sort of privacy and anonymity and do not allow the use of intonation, gestures, and feedback, participants in online support groups develop strategies to establish their legitimacy and their authority, such as referring to relevant experience or previous interactions, to signal that they are "entitled" to contribute and are trustworthy. For instance, blog authors are aware of their readers and adapt to them even in confessional blogs (Nardi, Schiano, Grumbrecht, & Swatz, 2004). Also, some bloggers have personal codes of ethics (e.g., never criticize friends) that they use to decide what goes into their blogs.

The enactment of the addressing mechanism online is made more difficult mainly because of the multiplicity of audiences in online communication. Many of our interviewees are working simultaneously with people in many different locations and contexts, with many audiences, sometimes known, sometimes unknown (e.g., when their messages are forwarded to unknown people without them being aware of it). A strategic planning director based in New York was working with people distributed in 22 countries, through several time zones, and across multiple organizations (the client, the different offices of his media agency, and the advertising agency). He kept reminding his teammates in New York how crucial it was to think of the recipient, of his or her context and culture, in order to build good working relationships and trust between the different team members. For this manager, despite our tendency to believe that the world is getting smaller and distance evanescing, distance still exists. In fact, another interviewee said that he thought that distance was even more important than before, exactly because of this apparent softening of the differences. Furthermore, some of the audiences are not intended or known at the time of the posting. In such contexts, one needs to tailor his or her message not only to multiple known others, but also to possible unknown others—including one's superiors, in case the message gets copied, or to the larger public, in case it gets posted in a public forum. In fact, several interviewees mentioned that when drafting an e-mail they were consciously thinking of possible future audiences, such as new people joining a project or a new manager who might not know the whole history and context. The boundary between private and public becomes porous in this context, and it may put a harsh brake on people's willingness to be open and emotional. Several interviewees said that before sending work-related e-mails they would carefully prune them of emotional expression, often using the drafting feature as a way to help them nuance the expression of their emotions and make sure that they address the other.

Reflecting

We have argued in Chapter 6 that the speed and increased synchronicity of online exchanges makes it difficult for people to reflect in an undisturbed way. When it comes to emotional expression, this may be seen as less problematic than when one tries to get the benefit of writing for knowledge creation. One could argue that the ability to pour one's heart out through the electronic device and then know that it can be read nearly immediately by the other can make writing closer to oral communication and, in that sense, more effective. However, the asynchronicity of writing is crucial to expressing emotions, as it allows us to take the time to express what we want to say and, even more, to become aware of what we feel, sometimes by going through various versions of the writing—e-mail or post. The installation *Take Care of Yourself* by the French conceptual artist Sophie Calle is an extreme example of this need to reflect in writing on one's emotional experiences. The installation was presented at the 2007 Venice Biennale and consisted of 107 outside interpretations of a "breakup" e-mail she received from her lover. She introduced the installation (which has also been published in a book) with these words:

> I received a dumping email. I did not know what to reply.
> It was as if it was not addressed to me.
> It ended up with these words: take care of yourself.
> I took this recommendation literally.
> I asked 107 women [...] to interpret the letter from a professional angle.
> To analyze, to comment [...] To understand for me. (Calle, 2007, p.1)

Instead of reflecting on her own, Calle invited 107 women with different professional vocabularies and perspectives (e.g., writer, forensic expert, lawyer, actress) to reflect for her and with her on the meaning of this breakup e-mail, which she read multiple times—an extreme example of objectifying—without being able to go beyond her grief. Through these 107 interpretations displaying multiple emotions—love, grief, sympathy, anger—she was able to articulate her emotions and make sense of them. Calle said, "After 1 month I felt better. There was no suffering. It worked. The project had replaced the man."[4] Hence, the installation was successful in providing catharsis and therapy.

Specifying

Specifying is very much intertwined with objectifying and addressing. Indeed, when writing down (objectifying) deep emotions, writers have to articulate specifically what they feel—for themselves as well as for the reader. Similarly, in order to address the recipient, writers specify the context, including their perceptions and feelings. While the speed of interactions may prevent the detailed specification of particular nuances of feeling and the search for the evocative word, other aspects of electronic communication—the use of emoticons (what in a letter would have been a tear or a drop of water on the page), the ability to use various fonts, the ease of inserting pictures and images—may help to specify emotions. Overall, the interviews we conducted and our experience suggest that the enactment of the specifying mechanism has more to do with the writer's motivation and practices than with technological features of the medium used. We discuss the enactment of the specifying mechanisms in contemporary practice on the basis of the following personal example.

I (Anne-Laure) recently went through an experience that shows the role that writing and, in particular, specifying, plays in expressing emotions.

I lost my 14-year-old cat, a full family member, who had cancer and had to be put to sleep. This was a very stressful and emotional event for my family, and for my two children (9 and 11) who had lived with Noa the cat all their lives. In fact, two months before the cat's death, the 11-year-old had written a poem about how much he loved his cat. As he had to briefly describe his poem, he wrote, "This poem is about my cat, Noa, who is a 14-year-old Siamese and has been in my life ever since I was born. In this poem I tried to describe moments with her and how lovely they are [...]." Writing was cathartic for the 9-year-old, who decided the day her cat passed away to start "a notebook of sad things" (her words). It opened with her experience of her cat's death: "the worst day in my life."

After the death, I wrote long e-mails to close friends, describing the feelings experienced by our family and friends: "And we are all here now—so sad, so empty [...] unable to believe [...] this is the sad reality of death...but it is hard." In most cases, I could not share my feelings with others orally, as they were complex, and I was feeling too emotional about the situation. Even among ourselves, we were not always able to articulate our feelings, and we sent each other e-mails for several days, saying how we felt, slowly realizing what this cat meant to each of us. In all these cases, writing down our feelings (objectifying) forced us to articulate them (specifying), thus making them, to a certain extent, manageable. These feelings, rather than overwhelming emotions that would leave us in tears, were articulated—specified—for ourselves and among ourselves. Like in the Calle installation, we were able to make sense for ourselves as well as together as family members.

Addressing the different recipients forced us to find words to describe feelings, which allowed us to also understand better the experience we were going through—beyond the loss of a loved pet, the experience of death, and the many fears and anxieties associated with it. In a time without e-mail, we might have sent letters or given a few phone calls (and prior to the phone, people sent telegrams for urgent messages). In our current time of multiple media, a few text messages (via mobile phone) were sent (which did not imply any relationship management involved in oral communication). Later on, long e-mails were sent. Electronic media allowed us to share our feelings and receive support from a large circle made of our extended family and friends.

While all messages were "customized" to each recipient, some paragraphs evoking some facts or general feelings were cut and pasted into messages to other friends. On the one hand, the first e-mail became an "original draft," and it allowed me to avoid going through the emotions involved in writing about my grief. On the other hand, one would argue that the ability to cut and paste (which was not possible with paper letters) allowed me to avoid the process of articulating and reflecting on my feelings that would happen each time I wrote. The object of the e-mails also varied and reflected the relationship I had with the person and how much that person knew about the cat. To my close friends I wrote a specific e-mail titled "Mademoiselle Noa, Princess of Siam, Princess with a long nose, has left us." To others, I mentioned the cat's death in the body of my e-mail.

The replies came fast—from a few hours to a day or two. Whether short or long, these replies were expressing the writers' feelings of empathy (e.g., "I could not spear a tear while reading your e-mail . . ."; "What a tough experience . . . And so much sadness. Lots of courage. . . .)." Some discussed the emotions expressed and felt and reflected on losing a loved one, as is expressed in this e-mail sent by a close and dear friend:

> Death is something in front of which we cannot do anything and in front of which we are left completely unprepared. Each of us is alone in front of this and has to deal with it. At the same time—and I can see it very well in your e-mail—it brings together, it creates links between those who are still alive, and who were link to this loved person or loved being.

This excerpt is a beautiful example of how emotions can be expressed in writing, when reflecting on the original e-mail. This story shows once again how writing allows us to turn our experience into words and provides us with a space to articulate and express our emotions, for ourselves and with others. Rather than freezing our emotions, writing, because it is a creative process, allowed us to go beyond the sheer pain of the feelings and to articulate and understand our emotions.

FROM PERSONAL LETTERS TO PERSONAL E-MAILS: HOW ONLINE WRITING CAN SUPPORT THE EXPRESSION OF EMOTIONS

While the case of the flame war in the open source community could be seen as a typical example of how online communication only leads to negative emotions, examples like Sophie Calle's installation or the story above suggest that online writing can also support the expression of emotions in a positive and supportive manner. These examples, as well as the story of the soldier and his wife discussed in Chapter 1, illustrate that rich personal correspondences still exist nowadays—if only we enact the mechanisms. Oftentimes people contrast e-mails that are seen as short and more informative with long, well-written letters, full of emotions and thoughts, such as the ones written by Virginia Woolf and discussed in Chapter 3. However, as noted in Chapter 6, there are many more genres of letters than the long, "well-written" letters to friends and families often associated with 18th- or 19th-century writing, and all these genres can be performed through e-mails. In particular, as seen with the example above, long personal e-mails, full of emotions, which would match the long, personal letters, can be written and exchanged among friends and families. In fact, both of us have been living abroad and have developed a correspondence with friends and families via e-mails. Similarly, our friends living abroad have developed e-mail correspondences with friends and family. Some of these e-mails have provided them with the opportunity to reflect and express emotions that could be more difficult to express orally, or that at least needed a specific context, which might not be found in the busyness of everyday life. In fact, the recipient of the e-mail did not need to be far away for the emotions to be expressed. What mattered was the asynchronous context of writing and how it supported the writing mechanisms. Being able to write at night to a friend, even if she lived a couple of blocks away, allowed friends to share emotions or personal reflections, discuss perspectives on topics of common interest, and maintain relationships.

Reflecting on her e-mail correspondence (kept in various folders in her e-mail account), one interviewee noted that her e-mails were very similar to letters she would have written 20 years ago. The major change brought by the technology was the increase in the speed of the interactions, which made it simpler as there was no more need to get an envelope, a stamp, and go to the post office or a mailbox. It was also easier to attach photos or videos. Sometimes, she would copy several people—for example, all her family members; this is not possible with letters. One might do that on Facebook, but she argued that the e-mail (even with several recipients) forced her to "tell a story" and address her prose to specific persons rather than just post photos. Some of her friends even started blogs with photos of their children and stories about their life.

All these examples from our interviewees, our personal lives, and the open source communities are exemplary proofs of how writing mechanisms can still be enacted with electronic media—e-mails, blogs, or posts in online forums. Consequently, these new media can be, just like the old letter-based world was, the theater of strong and nuanced emotions, a place in which strong bonds can form and be maintained between people.

CHAPTER

8

Creating Knowledge in Online Interactions

In an era where we can fly from one part of the world to another in hours rather than the days, weeks, or even months it took to travel in the time of Descartes, the Hudson's Bay Company, or even Einstein, and where video conferencing and other tools such as Skype and teleconferencing allow us to interact orally and sometimes with visual images across the globe, writing might seem a poor substitute to face-to-face interactions. Yet, a quick overview of the ways knowledge is created and shared today suggests that writing is far from being dead. Take, for example, online discussion forums that present a valuable venue for interaction among busy, working professionals and that are based only on written exchanges. In fact, these online forums are increasingly regarded as important venues for promoting learning across the boundaries of time, space, and formal organization (Ahuja & Carley, 1999; Butler, 2001; Gray & Tatar, 2004; Herring, 2004; Sproull & Faraj, 1995). Online forums provide propitious environments for knowledge sharing and learning as well as collaborating. As we saw in Chapter 7, the free and open source software developers have successfully developed complex software products relying almost exclusively on written exchanges.

Moreover, to rise to the innovation challenge posed by competitive economies, organizations are increasingly relying on distributed teams in order to tap into various pools of expertise and build upon contextual knowledge. Consequently, a lot of their interactions are taking place via writing—e-mails, written documents exchanged or posted on platforms, blogs, and so forth. Furthermore, companies, inspired by open source development, have in the last decade started to "ask" the crowd to provide solutions to their innovation problems through innovation-oriented crowdsourcing platforms, such as InnoCentive.com and TopCoder. In the science domain, a similar trend toward collaboration among people distributed around the world has emerged with citizen science efforts where volunteers, oftentimes with no specific scientific training, accomplish or manage research-related tasks such as observations or measurements. For example, in one of the most successful citizen science project, Galaxy Zoo, first launched in 2007 by astronomers and astrophysicists from the United Kingdom, participants are asked to help classify galaxies by studying images of them online and answering a standard set of questions about their features. Another famous example of distributed knowledge development through writing is Wikipedia, a free, Web-based, encyclopedia founded in 2001 and whose content is written collaboratively by volunteers, who can add and edit entries. It has about 10,000 regular contributors, consists of over 20 million articles, and is written in as many as 280 languages.[1] In open innovation platforms as in electronic mailing lists, such as Listserv, or in online forums, the main mode of communication is also writing.

In this chapter, we will review briefly the literature on knowledge development and online communication to highlight how our approach allows us to go beyond some of the contradictory findings we discussed in Chapter 2. We will then present the case of an open innovation platform, OpenIDEO, to illustrate how the mechanisms of writing are enacted by the members of OpenIDEO and allow them to share ideas and develop new concepts. We end the chapter with a discussion of the changes in the enactment of the mechanisms of writing brought about by the new media.

WRITING ONLINE AND KNOWLEDGE DEVELOPMENT

Despite the examples discussed in Chapter 4—as diverse as the development of new theories by a philosopher like Descartes or scientists like Cartan and Einstein, and the development of operational knowledge and organizational memory in a large distributed organization like the Hudson's Bay Company—most research on the capabilities of communication media to foster knowledge development seems to support the claim that collaboration requires proximity and cannot occur in distributed contexts (Olson, Teasley, Covi, & Olson, 2002). For example, McGrath and Hollingshead (1994) noted that groups using computer-mediated communication lack nonverbal cues and are never really sure why someone has not replied to a message, and that these limitations have a negative impact on collaboration. Similarly, Kraut, Galegher, and Egido (1988) reported that physical proximity, and the informal interactions it triggers, plays a key role in scientific research and collaboration. They therefore argue that the technology to support scientific collaboration should be as rich as possible (i.e., as close as possible to face-to-face interaction).

Recently though, these findings have been challenged by studies demonstrating how online communities and groups can excel at intellective and collaborative tasks. In spite of the many challenges such as geographical distance, status, culture, or identity that distributed teams relying mostly on computer-mediated communication confront (Hinds & Mortensen, 2005; Levina, 2005; Levina & Orlikowski, 2009; Metiu, 2006), studies have found that distributed collaborations can be successful. For example, Cummings, Schlosser, and Arrow (1995) found that computer-mediated groups produced essays with higher integrative complexity than those of face-to-face groups. Moreover, the open-source and cognitive science communities are two great examples of how computer-mediated communication can support idea production and exchange (Kogut & Metiu, 2001; Schunn, Crowley, & Okada, 2002). The rapid development of new IT-enabled forms of open innovation that rely on virtual organizing as a source of new ideas is another example of the possibility of creating and developing knowledge at a distance. Hence, studies on computer-mediated communication, idea sharing, and collaboration have produced conflicting results.

In fact, since its beginnings, online communication has been used to link a community of scientists at the U.S. Department of Defense (the ARPANET)—not unlike the networks of correspondents to which Descartes, Mersenne, Madame du Châtelet, or Einstein belonged. Since the days of the ARPANET, a lot more researchers and academics have found the modes of communication enabled by the Internet effective in exchanging information and building new knowledge and therefore quickly adopted them. Written online interactions offer similar possibilities to letters in terms of the richness of information exchange and development, yet with a different magnitude and speed. The ease with which e-mail and forum exchanges are stored, searched for, retrieved, and shared (partially or in their entirety) with others allows participants to easily find and categorize information using the traces left by the written threads. Many technology professionals praise e-mail for its permanence and the capability to archive messages (Yates & Orlikowski, 1992).

Yet, e-mails and other online text messages are crucial to knowledge sharing not only because they can be archived (as the output of the objectifying process) but also because of the objectifying process itself—as well as specifying, addressing and reflecting—enacted while writing these messages. Because they are written, computer-mediated technology such as e-mail and online newsgroups support the objectification process and distant collaboration. Thus, the objectifying and specifying mechanisms are what enabled software developers to develop complex products such as Linux and Apache almost exclusively via e-mails and forum exchanges (Kogut & Metiu, 2001; Moon & Sproull, 2002; von Hippel & Krogh, 2003). The same dimensions also help explain the way some types of experimental biology, such as the Human Genome Project, are conducted: several labs working on distinct parts of the problem, with one lab generating an intermediate product (such as a cell line) and then sending it to another lab for further work (Walsh & Bayma, 1996). As in letters, ideas are thus objectified, such that other scientists can access them frequently, reflect on them, and build their own contributions more carefully than in a face-to-face situation (Hiemstra, 1982). While letter writers do not always keep a copy of the letters they send, online communication offers writers the possibility to revisit their messages and posts. In that sense, online communication supports a higher level of objectifying, which enables collaborative scaffoldings.

Our approach of shifting the attention from the features of the media to the mode of communication helps explain some of the factors underlying the capability of computer-mediated communication, and of the Internet, to provide a propitious environment for distributed collaboration. While many studies note that online interactions can foster idea exchange, they explain it by the affordances of the medium, such as bandwidth, or archiving capacity (Rice & Love, 1987). Without dismissing the impact of such technological advances on the speed and scope of sharing, our perspective draws attention to the fundamental mechanisms of writing that are enacted during computer-mediated intellectual exchanges. We illustrate the importance of the writing mechanisms in the development of knowledge with a case study of OpenIDEO, an open innovation platform. The findings of our analysis of OpenIDEO communicative practices are confirmed by the interviews we did with professionals in various fields (media, advertising, publishing, architecture, consulting, coaching) and countries (France, United States, Singapore, United Kingdom).

OpenIDEO: Developing Creative Concepts About Social Issues

In this section, we will present a brief case study of an open innovation platform to illustrate how writing can support knowledge development in online contexts.[2] Indeed, open innovation and innovation-oriented crowdsourcing platforms such as InnoCentive.com and TopCoder (Archak, 2010; Lakhani & Jeppesen, 2007) have been publicized as the future of innovation. These platforms allow thousands of individuals to contribute ideas and develop products or concepts by replying to publicly posted challenges involving problems from fields as varied as bioscience, engineering, chemistry, business, and social innovation.

OpenIDEO: An Open Innovation Platform

OpenIDEO, a Web-based platform for open innovation in the field of social innovation launched in August 2010 by IDEO is a perfect example of this attempt to develop creative yet implementable ideas from a vast community of distributed participants. IDEO, founded in 1991, is a world-leading design consultancy with offices in Palo Alto, San Francisco, London, Boston, and Shanghai, among other places. Consistently ranked as one of the most innovative companies in the world, IDEO is famous for its method of innovation based on a design process and intense cross-disciplinary project work and has won many international design awards.

In September 2011 there were more than 18,000 OpenIDEO members located all around the world. They are invited—completely voluntarily and with no prizes or awards—to participate in challenges set by organizations, public and private, such as Water and Sanitation for the Urban Poor (WSUP), Oxfam, the Haas Center for Public Service at Stanford, Grameen Creative Lab, Unilever, or Nokia. These challenges are all related to social innovation (i.e., seek "to create transformational change in underserved, underrepresented, and disadvantaged communities worldwide").[3] Examples of challenges are: How can we improve sanitation and better manage human waste in low-income urban communities (sponsored by WSUP and Unilever)?; How might we improve maternal health with mobile technologies for low-income countries (sponsored by Oxfam and Nokia)?; How might we better connect food production and consumption (sponsored by the Government of Queensland, Australia)?; and How might we increase the number of registered bone marrow donors to help save lives (sponsored by the Haas center for Public Service, Stanford University)?

The OpenIDEO platform mimics the IDEO process, a human-centered, design-based approach, where observing, brainstorming, and prototyping are key activities. The IDEO team is involved in framing the challenge, encouraging the conversation, and prototyping. The OpenIDEO platform structures each challenge around three phases: inspiration, where members are invited to do some research on the challenge and share their ideas; conception, where OpenIDEATORS build upon the ideas collected and shared in the previous phase, and develop and articulate concepts that could have an impact; and evaluation, where 20 concepts are selected by the OpenIDEO team with the challenge sponsors, taking into account the feedback of the community members, and where OpenIDEATORS are asked to evaluate them. OpenIDEATORS can contribute in many different ways: by posting observations, inspirational projects, sketches and photos, or simply by giving feedback and evaluating other members' concepts. Feedback through comments is invited at every step of the process. At the end, the 10 strongest concepts are selected by a jury composed of the sponsors of the challenge and the OpenIDEO team, using the evaluations provided by OpenIDEATORS. Yet, all concepts are maintained on the platform, ready to be shared, reused, or recombined to solve a future challenge. The hope is that some of these concepts will be put in action either by the sponsors or by other organizations or some of the participants on the platform.

OpenIDEO aims to support collaborative creation of ideas and is recognized as such by the participants, as suggested by this blogger who wrote on his blog the day of the launch of the OpenIDEO platform:

> I'm really inspired by the site because it realizes a very important point: ideas aren't scarce. Now it's not about gathering tons of those ideas just to collect them, it's about creating a framework where ideas can inspire each other. I think the smart cookies behind OpenIDEO have nailed this in the site design. The experience basically creates like the largest, most unorthodox design team in the world thinking, submitting, and churning on some really big problems.[4]

OpenIDEO is a great illustration of collaborative knowledge creation through writing with participants posting inspirations, commenting, elaborating, and modifying their concepts to develop solutions for challenges posted on the platform. We did a qualitative discourse analysis of public posts and comments for four challenges: Maternal Health, Food Production and Consumption, Social Business in Caldas, and OpenIDEO Impact. Our analysis shows that knowledge creation and sharing on the OpenIDEO platform takes place mostly through writing. In fact, the practices enacted by the participants on the OpenIDEO platform, also called OpenIDEATORS, are quite similar to the practices enacted by Cartan and Einstein—generating, debating, and articulating ideas—while developing a theory. In the section below, we present three communicative practices unveiled by our analysis: generating ideas, giving feedback, and articulating actionable concepts. All four writing mechanisms—objectifying, addressing, specifying, and reflecting—are central in generating ideas, giving feedback, and articulating concepts in a collaborative and distributed manner on the platform.

Generating Ideas

The OpenIDEO platform aims to generate ideas among participants from various backgrounds and in different geographies in order to solve problems related to social issues. The generation of ideas takes place all along the challenge, and right from the beginning in the inspiration phase where people share various ideas, whether related or not to the challenge. For example, in the Maternal Health challenge, OpenIDEATORS posted existing products, services, and campaigns that improved health (maternal and otherwise) and shared their positive or negative experiences with maternal health along with some innovative uses of mobile technology that could be inspiring.

In this process of idea generation, the objectifying mechanism allowed people to share ideas and create a repository onto which others could build. Participants are encouraged and often refer to others' inspirations and concepts, as illustrated by this OpenIDEATOR's post in the Food Production and Consumption challenge:

> Drawing on a number of incredible inspirations (including DC Central Kitchen, La Cocina, San Francisco Foodshed Project, and SF Foodcart Project) this model takes things one step further by creating a community driven, but closed loop system that simultaneously creates both the supply and demand needed for it to be sustainable. (April 28, 2011)

In this post, the OpenIDEATOR is articulating his own idea but also referring to others' ideas, which have already been objectified and posted on the platform. Hence, objectifying allows him to generate ideas, by writing them down as well as by being able to reflect and refer to others' ideas, made available to all on the platform. Objectifying is also enacted when users post external links to expand on their ideas or give multiple examples in order to explain their point of view.

Generating ideas is more than simply sharing ideas that have been found; it also involves discussing, commenting and collaborating. In fact, community members comment a lot, leading the person who posted the inspiration or the concept to specify her or his ideas and sometimes to generate new related ideas. For example, in the Maternal Health challenge, Sofia posted an inspiration about JAMA, a clean-birth kit, which can help prevent many diseases.[5] Jeanne posted a comment providing general feedback and asking for specification:

> Great example for a simple and very practical solution, thanks Sofia. I wonder if they have some (cartoon-like?) visual guide, to help the users. And would it be possible, to replace some of the tools with local products? (February 17, 2011)

Sofia replied that while she had not heard about a practical guide, she knew that the distribution efforts were complemented with education, and through partnerships with local organizations to provide training to the women. While in this case Sofia specifies by providing extra information about the clean birth kit, in other cases, comments also led to generating more ideas. For example in a challenge about how to improve the social impact of OpenIDEO (sponsored by OpenIDEO for its first anniversary), one of the concepts was a phone application that would make the platform accessible with a smartphone—"an easy-to-use app that we can use to jot down our bursts of inspiration and quickly share it with those in our network." Many OpenIDEATORS commented on this concept, making suggestions for adding more features; here is one example:

> If we could incorporate on-the-go inspiration everyday & anytime, maybe have an 'image bank' of this kind of contribution in the app. Kind of like CNN's iReporter function on their app where any citizen can report a special event they are witnessing to CNN. (September 8, 2011)

The OpenIDEATOR who developed the concept replied: "Thank you. And you make a good point about CNN's iReporter. It didn't strike me at all. Yes, I guess it should be kind of like that!" (September 8, 2011).

In this case, the analogy with an existing system triggered reflection and knowledge advancement ("it didn't strike me at all"), and a stronger possibility of implementation in the final application. In this reply, we can also notice the reflecting process, which leads to the recognition that this idea could improve the concept.

To make their ideas more easily intelligible to and interpretable by the distributed community, posters often provide background information for their comment such as the following comment on the app concept:

> Just a counter-point thought. With a simple mobile phone and a dial-up internet connection I believe I represent a proportion of those we are trying to reach. I (and people like me) could act as weather vanes for platforms accessibility—thinking of which accessibility for those with vision, hearing and other impairments bears consideration too. (September 6, 2011)

Generating ideas is at the core of the activities taking place on the OpenIDEO platform, and it happens mainly through writing, complemented with a few visuals—images, sketches, or videos. Writing's mechanisms, in particular objectifying and specifying, are central to the generation of ideas by OpenIDEATORS.

Giving Feedback

For someone who goes on OpenIDEO platform for the first time, as a lurker or as a poster, the quality and style of feedback is remarkable. OpenIDEO members give each other a lot of detailed feedback—commenting, asking for clarification, making suggestions, and providing ideas—all in a very friendly, enthusiastic, and positive tone. Most of the time, they refer to each other's names, congratulate, and thank one another. "Flaming" (see Chapter 7) is completely out of the question here, and members simply do not post even slightly negative comments. The process of getting feedback is crucial to the evolution of ideas and is encouraged through the different phases of the platform, the comment feature, but also through the possibility to update a post. The different phases, from inspiration to evaluation, provide clear steps for developing one's ideas and many opportunities for feedback. This structure invites participants to think of idea generation and concept articulation as an ongoing collaborative process.

Commenting is at the core of the giving feedback practice. Comments are frequent and they usually are personal with a lot of relationship management: "Hi," "Dear," and "Thanks." Comments are "threaded" (i.e., they are presented in a way that makes it visually clear who is replying to whom). This creates a strong sense of dialogue that reinforces the addressing communicative practices. Moreover, in many cases, the feedback is taken seriously and often inspires OpenIDEATORS to modify their initial concept. Often, they refer to people who influenced them in developing their concept, sometime even referring to a virtual team at the end of their post. For example, one conceptor, in a challenge on developing social business in Caldas (Columbia), quoted specifically the different commentators and modified his concept based on their comments:

> You could also imagine that the truck could even provide ways for local crafts to show and sell their stuff. Imagine women weaving in a village. They could give one or two samples of their work which could be exhibited in the truck. Thanks to Jeanne W. for this great addition. (June 28, 2011)

This referencing—grounded on the objectifying and addressing mechanisms—is supported by the built-upon feature of the platform. This feature offers posters the possibility to connect their ideas with others' ideas by a simple drag-and-drop action. The connected idea is presented on the side, making visible to everyone the connections between different ideas. The importance of connecting ideas and recognizing others' feedback in one's concepting process is also symbolized by the possibility to form a virtual team. Often, OpenIDEATORS mention as part of their virtual team other OpenIDEATORS who commented and gave feedback on their ideas.

Replies from the poster in the comments' thread as well as changes in the concept also illustrate the reflecting process. In fact, one of the community managers highlighted the importance of the reflecting mechanism and encouraged it in one of her comments to an OpenIDEATOR who provided feedback on a concept—feedback that was implemented in the revised version of the concept:

> Mateo—this kind of discussion is really great in the Concepting phase and you've brought a great perspective to the conversation of Mary's great idea—which can only get better from your input!

Giving feedback is therefore a key practice to the OpenIDEO process, which is intrinsically collaborative and evolving. Feedback is elaborated and shared through writing and objectifying, addressing, specifying, and reflecting are all supporting the feedback process.

Articulating Actionable Concepts

One major goal of the OpenIDEO process is the creation of concepts that provide solutions to the challenge at hand and that would be implemented by the sponsors of the challenge or by other people, whether the OpenIDEATORS or other organizations engaged in similar issues. Articulating concepts based on the ideas generated in the first phase and on the feedback received is the ultimate aim of the activities on the platform.

The articulating practice is supported by the OpenIDEO team through various features and guidelines. For example, the "built-upon" function makes the referring very easy and provides not only recognition to others but also background information to other OpenIDEATORS. This built-upon feature is a way to enact the objectifying mechanism by referring to other ideas or concepts, which have been already posted.

When posting a concept, participants have a list of specific points and questions to address in their concept, such as these main questions in the Social Business challenge sponsored by Grameen Creative Lab: How do you envision this idea making money?; How does this idea create social impact, particularly around improving health?; and How does this idea add social value at every step of the process? These questions invite the OpenIDEATORS to be as specific as possible while writing and articulating their concepts.

Specifying is also encouraged by the community managers as well as by the comments of other participants. If the OpenIDEATOR has not provided enough background, the platform provides features that ask for quick clarification. Hence, addressing here when not completely enacted—and it is not always feasible to address all possible questions as one is writing to a community of several thousands people, located in various geographies and with diverse background—can easily be extended. This is illustrated by the following two comments in the Food Production and Consumption challenge:

> Hi Sophia, I think this business idea is over my head. I see many responses that point out the creativity and potential benefits of the idea, but I don't really understand. How would a person (like myself, as a consumer) get involved in this project and how would it make food more accessible to low-income regions? (April 28, 2011)

> Hi Mike, thanks for asking for direct clarity. As an end consumer, you would get involved in this project by purchasing food from the variety of food entrepreneurs that are a part of the kitchen—probably a majority of them operating food trucks. (April 28, 2011)

As illustrated in the excerpts above, comments are most of the time addressed to specific members. In many cases, people refer to their own expertise or background to ask for clarification or make suggestions. For example, a member in the Food challenge commented based on his own knowledge of markets in Europe. Also, in the Maternal Health challenge, a member made a comment on a concept referring to her own experience working in nonprofits in South Africa. In some cases, as in the Food challenge, the OpenIDEATOR clarifies her or his idea in the comment. In other cases, the OpenIDEATOR will update her or his concept—sometimes highlighting the changes by a mention of "Updated."

Articulating ideas into an implementable concept is an objective of the OpenIDEO platform, as mentioned on the website: "All concepts generated are shareable, remix-able, and reusable in a similar way to 'creative commons'. The hope is that some of these concepts will become reality outside of OpenIDEO.com."[6] The importance of making an impact and making these concepts potentially implementable outside of Open IDEO.com is symbolized by the introduction of a new phase, the refining phase[7]—which is strongly grounded in the specifying and reflecting mechanisms of writing. The refining phase, which takes place between the conception and the evaluation phases, short-lists the final 20 or so concepts, and the representatives of the organizations who posted the challenge raise questions about the possibility of implementing the concept. The short-listing is done by IDEO and the participating organizations, based on the applauds and comments of the OpenIDEO members as well as by using evaluation criteria (such as feasibility). When the introduction of the refinement phase was announced, a community member commented:

> Wow Niki this refinement phase is a great addition! Really facilitates collaboration and the fleshing out of concepts to increase realization potential. And I love the honorable mentions feature, as it recognizes that ideas often emerge from multiple sources. Congrats to everyone on coming up with some amazing ideas for this challenge, and look forward to collaborating with all of you on pushing our concepts to the next level! (March 25, 2011)

In this comment, the participant highlighted the importance of specifying—"fleshing out"—and how the platform supports collaboration. The mechanisms of writing and the asynchronous nature of the collaboration are crucial here. Indeed, through objectifying, participants are able to articulate ideas and share them for others to reflect on them, asking for clarification when needed, and specifying them in response.

The refining phase allows conceptors to reflect on the feedback and develop their ideas based on the feedback they receive. It also allows members of the community to focus on the 20 concepts that have been selected by the OpenIDEO team and the challenge sponsors based on the applauds of the members of the community. Like generating ideas and giving feedback, articulating actionable concepts is a practice that takes place through writing and relies heavily on the enactment of all four of writing's mechanisms. In many ways the platform—through its features and guidelines—invites and reminds OpenIDEATORS to enact these mechanisms, which are crucial to successfully generate ideas, give feedback, and articulate concepts.

Knowledge Development at OpenIDEO

OpenIDEO is a perfect example of distributed collaboration, where members of the community, who, for the most part, have never met, take time to think about challenges related to social issues, to research the problem and share their findings with others, and then develop creative concepts that can be implemented. Some in fact are in the implementation phase, such as a home toilet prototyped in Ghana (in the Sanitation challenge), FlavourCrusador, an app to connect farmers and food-lovers to create the future of food (in the Food Consumption and Food Production challenge), or the "Social Change in a Box" concept, a website providing informational materials, promotional files, and more, all easily sharable, understandable, and adaptable to many situations (in the Bone Marrow Donor challenge).

The OpenIDEO case exemplifies the new forms of organizing and collaborating enabled by communication and information technology. As it is common in online collaboration, most of the interactions took place in writing. All four mechanisms of writing—objectifying, addressing, reflecting, and specifying—were enacted by the participants as they created, developed, and shared knowledge. While photos and videos are used to illustrate ideas and provide more context and details, the core of the creative process takes place through writing.

The knowledge development process at OpenIDEO illustrates clearly that the objectifying process does not freeze the ideas as people keep evolving (and are encouraged to do so by the OpenIDEO team) and updating their concepts based on the feedback of others. Also, the comments sections act as a two-channel communication between the poster and the other members. Last, the platform itself by providing different phases suggests a process, which is never completely locked and always ongoing. Furthermore, new media make it easy to continuously and significantly expand the number of people working on a problem. In all phases, the OpenIDEO members are given the options to share a post that they read or wrote within their social networks by linking them to Facebook, LinkedIn, and Twitter. These features of the new media provide an opportunity to reach a greater number of people working on a particular problem, and the individual can work with the people they personally know or professionally are cohesive with. Thus, using the power of a sense of community, responsibility, and the provision of intangible rewards, OpenIDEO has created a productive public platform in which members share and build knowledge. In the following section, based on our interviews, we unpack the writing mechanisms in developing knowledge in different organizational contexts such as media and advertising agencies, architecture firms, publishing companies, and management consultancies.

DEVELOPING KNOWLEDGE IN ONLINE CONTEXTS

Our findings regarding the role of writing for knowledge development with the OpenIDEO case are confirmed by the interviews we did. Indeed, all our interviewees insisted on the importance of writing in their work practices and in the development of ideas and content. For example, the founder and director of an architecture firm explained how in the last 10 years writing has become increasingly important to accompany the blueprints and drawings, to a point where he sends his employees to writing classes. While writing could be done with paper and pen, most of our interviewees' writing was done with electronic document and presentation software and via e-mail. It is important to note that the people we interviewed were involved in different fields and practices, and that the nature of their work influenced the style and content of writing. Yet, despite these variations in style and content, writing was always acknowledged as playing a role in developing knowledge. In line with the findings of recent studies (Barley, Meyerson, & Grodal, 2011; Mazmanian, Orlikowski, & Yates, 2006), the interviewees also acknowledged their addiction to e-mail as well as the symbolic role of e-mails and of their mobile devices, describing various practices to manage their stress and their status through e-mail.

While stressing the importance of writing, our interviewees also recognized the effects on their writing practices by the changes in the technology infrastructure and features. Yet, when introduced to the four mechanisms, all our interviewees agreed that they were essential in allowing them to share and develop ideas with distributed and collocated collaborators. In the section below, we unpack the four mechanisms of writing and their role in knowledge development, as described by our interviewees.

How the Four Mechanisms of Writing Enable Knowledge Creation With New Media

Objectifying

Objectifying or writing down was recognized by all interviewees as crucial to knowledge creation and sharing. Indeed, our interviewees insisted on the importance of writing for their own thinking process. Several drafted e-mails that they used as ways to articulate their thinking. One interviewee, a marketing director in a service company, explained how drafting her thoughts in an e-mail to herself rather than in a document put her in the mode of thinking of the person she was writing to and thus led her to be more specific. Another interviewee also talked about how she finds "the process of writing very clarifying, it helps figure out what you mean. I notice now more and more with e-mails I do a first draft and a second draft, I realize parts of it aren't clear." An interviewee referred explicitly to the way in which the writing process helped the writer develop and refine ideas as well as come up with new ideas. Studies of professionals in law, finance, advertising, and design also highlight the importance of writing and taking notes, as it helps them to construct and organize their thoughts (Kidd, 1994).

Knowledge development in some contexts would be done in the core of the e-mail and the attachments. In other cases, platforms similar to Dropbox or Google sites offer people the possibility to share and/or collaborate on a document. A vice president in a publishing company said that her company uses a platform where employees can upload different versions of documents. Therefore, people working on a project would upload the latest version and send a short e-mail to others working on the project, informing them of the new upload and briefly describing the revisions done.

While objectifying is important to all our interviewees, this is often not done on smart phones or other mobile devices but on laptops or desktops, because of the materiality of the interface (too small, keyboard difficult to use), the context of use (in a train, on a subway, in a meeting), and the access to other information (access to previous versions of the document or to other documents). Our respondents concurred in the notion that the computer is more conducive to long, elaborate, written communication than other devices such as smart phones. One senior manager in a technology company said: "I write only on my computer. See, I like writing, making sentences; for me replying to e-mail on a small device is very frustrating."

Addressing

By increasing the number of people that can be accessed (with the ability to copy many people on an e-mail) and the speed of exchange, the new media make people aware of the importance of addressing. Clearly, the increase in the potential number and diversity of the participants is one of the main characteristics of current collaboration technologies. Theoretically, everyone with an Internet connection can be an OpenIDEO member, a developer of free and open source software, or a contributor to Wikipedia. In OpenIDEO, members can also post their social media pages (e.g., Facebook or Twitter), thereby increasing exponentially the potential reach of their ideas and the possibilities for knowledgeable feedback.

At the same time, this increase in the number and diversity of collaborators also means that one has to think about ways to address this large and diverse audience. Yet, often because of the increased bandwidth and the speed of interactions, we tend to forget about the geographical and cultural distance that often separates collaborators. As both asynchronicity and perceived distance diminish, people often do not address an audience. They think of the e-mail recipients as being "in the same room" and sharing the same context. In fact, this is not the case and therefore can create a lot of misunderstandings. One of our interviewees, a strategic planning director in a media agency, insisted on the importance of thinking of your audience when crafting an e-mail that is sent to 22 different markets. Even when writing to only one market, he explained how he insisted that the people in his team keep in mind their audience—whether it was someone from the office in Germany, Spain, Singapore, or Australia. This was key when he was sending a request for information that he would then need to analyze and integrate in a strategy analysis or brief so that he could get the right information and good quality data. It was also important to tailor the tone and style in order to build a relationship and increase the chance of a response (as the different markets were not reporting directly to him and had no incentives to reply to him). The vice president of advertising sales in a global medical and health publishing house also insisted on how many people tend to think—wrongly, he argued—that because of the world becoming more global, cultural differences have been reduced. Beyond the cultural differences, one of our interviewees insisted on how while writing he always tried to think of how someone who did not know the context, or would come in the conversation later on, would understand his e-mail. This was crucial in sharing knowledge but also in being able to collaborate on a project and develop a shared understanding. This led him to craft his e-mail while consciously enacting the addressing mechanism.

Addressing is key for knowledge development as it allows the writer to clarify the background information needed by the recipients of his or her e-mail or posts. Yet, while writing an e-mail to a single individual or a small group, addressing is very similar to what the Hudson's Bay Company commanders or Einstein and Cartan were doing in their letters; it becomes more difficult when you post on an online platform or blog with potentially thousands of readers, or even when you work in distributed environments with more than two or three collaborators, who are often based in various countries or belong to different organizations (with different needs, objectives, and understandings). Moreover, the speed of interactions, by reducing the asynchronicity originally inherent to written interactions, leads us to become less cognizant of the need to address an audience.

Reflecting

The great gain offered by the new media in terms of quick access and speed is balanced by the pressure to reply quickly and thus the risk (and unfortunately the tendency) of doing so without sufficient reflection. Besides the speed of interactions, the interruptions created by new e-mails, often signaled visually or aurally, the plethora of communication means (e-mail, SMS, Facebook, Twitter), and the context of use (in a meeting, in the subway, at the airport, or in a taxi) lead to a lack of reflecting. Or, as we have seen in the first part of the book, reflecting is central to knowledge development. In fact, many of our interviewees expressed their frustration at the difficulty of enacting this mechanism. While they repeatedly stated that writing "frees reflectivity," they also complained that in corporations the expectation is "often to do things quickly, so there's a high response rate to e-mails but a low thought rate. Like getting the e-mail handled versus thinking about what the intention of the communication is"—as one of our interviewees, a freelance consultant, stated.

Some features, such as cut and paste, also dramatically hinder the reflecting process. For example, a strategic planner in an advertising agency talked about how she noticed that her colleagues and clients tended to be less analytical in the way they developed and presented their ideas. She explained this by the fact that people were increasingly cutting and pasting previous documents into new documents and presentations. Thus, they did not take the time to reflect on the problem and the possible solutions and therefore were often less creative. The CEO of an innovation consultancy told us similarly that she noticed how the speed of exchange of documents—often created by cutting and pasting previous documents—led people at clients' organizations to sometimes get bored with an idea and stop reflecting on it before it was even fully developed or officially discussed.

Of course, as we saw above, reflecting can be enacted with some of the new media. In fact, all our interviewees noted that reflecting was a key mechanism to writing and crucial to their ability to work and develop new concepts. The richness and quality of the posts and feedback on OpenIDEO also indicate the importance of reflecting. Reflecting is also key in allowing open source software developers to collaborate and develop new code. Yet, many of our interviewees also noted that reflecting was more difficult to enact in today's world, partly because of the speed of interactions and the pressure to reply quickly that grew from it, partly because of some of the technology—mobile devices—they use.

Specifying

Specifying is intrinsically related to objectifying, reflecting, and addressing. Indeed specifying is only possible because of the process of writing down ideas, of the effort to clarify things for an absent audience, and of the reflecting process that makes us realize missing points or gaps in our arguments. Hence, as all the other three mechanisms can be enacted with the new media, specifying can be enacted, too. However, it is exposed to the same limitations as those highlighted above. Thus, while objectifying and reflecting are difficult with a small, hand-held device used in a public space, specifying is similarly challenging. While pressure to reply quickly makes addressing and reflecting difficult, specifying is affected in a similar fashion.

Moreover, while it is still possible to write long Word documents or e-mails explaining ideas and developing arguments, in many organizations, people are told that people won't read more than two pages or three slides. Yet, as one of our informants noted, "How can you develop an innovative strategy in two or three slides max?" This is difficult, often impossible. Similarly, as Tufte (2006) showed in his analysis of the challenger accident, the NASA management could not see the problem because they were presented with a PowerPoint presentation rather than a written scientific report, which was the only way to specify all the elements at play and their interactions. (For an overview of the impact of PowerPoint and presentation software use on knowledge sharing and development, see sidebar PowerPoint Slides: Support to Oral Presentations or Written Documents?)

POWERPOINT SLIDES: SUPPORT TO ORAL PRESENTATIONS OR WRITTEN DOCUMENTS?

PowerPoint presentations, which are present in all corporations, government bureaucracies, and increasingly in universities and schools, represent a striking example of the blurring boundaries between oral and written modes that have been accompanying the emergence of new communication technologies. While PowerPoint presentations can be described as "written," they were conceived as a tool to support oral presentations, and they lack the analytical dimension that one can achieve when writing with sentences and paragraphs (Shaw, Brown, & Bromiley, 1998; Tufte, 2006). Yet, PowerPoint presentations are increasingly used as written documents—PowerPoint texts (Yates & Orlikowski, 2007) become deliverables, printed or uploaded in Web-based presentations, and shared and reused in the process of strategy and decision making (Kaplan, 2011; Tufte, 2006; Yannis, 2008). The consequences for organizational processes are serious: "As electronic or paper renditions of PowerPoint texts become the only record of major activities, comprehension is reduced and organizational memory deteriorates" (Yates & Orlikowski, 2007, p. 88).

Edward Tufte, Emeritus Professor of Political Science, Computer Science and Statistics, and Graphic Design at Yale University, wrote a famous pamphlet, "The Cognitive Style of PowerPoint," criticizing PowerPoint and its cognitive style, such as the organizing of every type of content in an extremely hierarchical, single-path structure with bullet-point lists and the breaking up of narrative and data in short fragments. Tufte warns us that while "many true statements are too long to fit on a PP slide, [. . .] this does not mean we should abbreviate the truth to make the words fit" (Tufte, 2006, p. 4). Other scholars, such as Shaw, Brown, and Bromiley (1998), showed in their study of business planning practices that bullet-point lists were too generic and did not specify relationships, thus concealing the analytic structure of the argument. What these studies highlight is how in fact PowerPoint does not support the writing mechanisms. As Tufte claims, this tools reduces our analytical thinking, and to a certain extent, even dilutes thinking. In some cases, the implications can be dramatic, as in the case of the Columbia shuttle accident in January 2003. Tufte shows that the NASA officials failed to see the problem because, instead of reading the engineering reports, they based their judgment on PowerPoint presentations, which were too generic and misleading in terms of the analysis (see Tufte, 2006, p. 7 for a longer analysis and additional references).

Tufte's criticism is shared by many who perceive PowerPoint slides as a poor replacement for developing an argument, and consider that a good PowerPoint presentation requires the articulation of an argument, a story. Hence, the CEO of an innovation consultancy in New York argued that she used PowerPoint to present, organize, and articulate a story that she had written before. Yet, she noted that many of the company's clients used PowerPoint without knowing what they were using it for, without thinking about the story beforehand. Another interviewee, a VP in a multinational technology firm, prefers well-written memos to PowerPoint presentations in which the linearity of the text and the attending logic are missing. He states: "People don't bother figuring out the essential. We're downloading our brains instead of conveying a message." All these interviewees are consistent with Tufte's criticism of PowerPoint in business settings, and his point that because only 40 words can be put on a slide creates the multiplication of slides, which are presented to an audience that has a hard time understanding the context and evaluating the relationships between the different pieces of information. To conclude, the features of the slides, and the cognitive style they afford, as well as the way people use them, are not conducive to the enactment of the writing's powers.

USING NEW MEDIA TO ENACT WRITING MECHANISMS IN DISTRIBUTED COLLABORATION AND KNOWLEDGE WORK

All our interviewees agreed on the importance of writing for knowledge development, mentioning two main reasons for this. The first is the uninterrupted train of thought that writing supports, which has the ability to take the writer to a place he or she wouldn't have envisioned otherwise; as one of our interviewees, a partner in a large consulting firm, stated, "Writing allows communication at levels that speech cannot, maybe because it permits the development of a longer stream of logic without the opportunity of rebuttal." This creative power of writing can be enacted in individual writing (as one interviewee noted, "The analysis associated with articulating a message in writing leads to new ideas") or through the collaborative process, such as the one in place at OpenIDEO.

The other main reason for the importance of writing is that one can choose words carefully, precise and evocative words that have the power not only to express new ideas but also to stimulate the mind and trigger complex responses (because of their different, rich interpretations or emotional associations, for example). Thus, writing allows the writer to be more specific and clear. For example, when one writes, one can eliminate inconsistencies because one can choose between words with a reflective selectivity (Goody, 1977). Unfortunately for today's workers, continuous interruptions and pressures for quick responses are not conducive to enacting the creative power of writing. As one of the interviewees said, "Very well-done writing takes much more time than does superficial oral communication." Yet, despite the increased difficulty in being able to do "well-done writing," many interviewees talked about how important it was for them to express themselves well in writing. An independent consultant said, "Some e-mails require a lot of energy because they're project related and I know they may be passed on to someone else. I want to make sure it's well written. E-mail is not something I take lightly."

Clearly, some of the media affordances (e.g., hand held, mobile), the increased bandwidth and access, and the resulting increase in the speed of interactions can constrain the ability we have to enact the writing mechanisms in the process of knowledge sharing and development. At the same time, new media offer us other options, such as the ability to easily draft and then edit our drafts, which makes it easier for people to go back to previous documents and edit and revise them knowing they will not have to start from scratch. New media also afford us the ability to connect to many more people, by copying several people on an e-mail while working on a project or by allowing us to access communities possessing a vast variety of skills and expertise. Even the cut-and-paste feature can be effective if used well.

Thus, while new media can sometimes make the enactment of writing more complex, our analysis of the OpenIDEO interactions and interviews show that all mechanisms involved in knowledge development can be enacted with new media. Technology and new media are never standalone entities; their interpretation and enactment are shaped not only by their affordances but also by the socially and materially situated practices associated with them.

CHAPTER

9

The Role of Writing in Developing a Sense of We-ness in Online Communities

The idea that writing is at the center of the scientific enterprise and the development of bureaucratic organizing is quite intuitive and well accepted. For example, letter writing allowed companies such as the Hudson's Bay Company (HBC) to grow as successful distributed organizations, and online writing supported the development of complex knowledge such as the Linux operating system. That writing can allow us to express deep emotions that we might have a hard time articulating in a face-to-face context is also gaining acceptance; think of Virginia Woolf's beautiful letters but also of the emotions expressed in the correspondence between the post commanders of HBC in Canada and the managers in the London headquarters, or in the exchanges among software developers scattered around the world.

These two powers of writing—for expressing emotions and developing knowledge—go a long way in explaining the flourishing of knowledge- and affect-based communities that communicate mostly through writing. Indeed, as we have seen in Chapter 5, communities of scientists and intellectuals emerged at the same time as the rise of the great systems of writing (Collins, 1998), and for most of their existence, these intellectual communities, such as the Republic of Letters in the 17th and 18th centuries, have communicated via writing and especially through letters. Letters were also used to support the expression of emotions. During the

First World War, war godmothers wrote to soldiers in the trenches, providing emotional support and meaning to their everyday lives.[1]

Today, the advent of information technology powerfully impacts the expanse of both knowledge and affect-based communities and the speed with which bonds can form. In the case of knowledge-based communities, speed and reach (including easier access for the outsiders) are the main changes. Hence, nowadays, a scientist or software developer (really, anyone) can post a question to "everyone" (who has an Internet connection) and get an answer. In the case of support communities, the change is, we argue, even bigger. Indeed, historically such communities had more difficulties arising in the absence of collocation. While socioemotional bonds for people at distance (Chayko, 2008) have always existed, as illustrated by "imagined communities" such as nations and diasporas (Anderson, 1983), technology is offering the possibility to extend the reach and strength of our bonds. Cancer support groups provide social and emotional support to their members, allowing them to have access to information, to share their experiences and fears, and to develop relationships regardless of physical location. In a similar fashion, the Vietnamese diaspora created by the war was able to build a global community, a "virtual Vietnam," despite the geographic distance (Lam, 1998). Nowadays, these emotionally bounded communities are forming online, and people are participating and feeling part of a shared group and developing emotional attachments. Therefore, some scholars praise online communities for bringing together individuals, regardless of their physical location, and providing support to people who otherwise wouldn't have had access to a group interested in similar things, thus arguing that online communities are *real* communities (Rheingold, 1993). Still, others consider online communities as less real and less strong in terms of human bonds than traditional, face-to-face communities (Kraut, Kiesler, Mukhopadhyay, Scherlis, & Patterson, 1998; Nie, Hillygus, & Erbring, 2003; Stoll, 1995).

In this chapter, we briefly review the debates about the possibility of creating a sense of *we*-ness, or shared identity, in online communities. We then analyze a particular online community, KM Forum, an online forum dedicated to knowledge management, and show how the mechanisms of writing help create a sense of *we*-ness among the members. We end with a discussion about the role of writing in creating and sustaining online communities.

BOWLING ALONE OR DIGITAL HABITATS: CONTRADICTORY FINDINGS ON HOW TECHNOLOGY INFLUENCES COMMUNITIES

Studies of online communities are always, at least implicitly, referring to issues related to community building and belongingness. Indeed, whether their aim is knowledge development or emotional support, both types of communities involve a form of shared identity. Also, the studies specifically focusing on the development of shared identity—in other words, a sense of *we*-ness—by participants in online communities show that such a sense of community is crucial because people are becoming increasingly involved in multiple distributed groups and cross-organizational arrangements, such as communities of practice (e.g., Fayard & DeSanctis, 2008; Star & Ruhleder, 1996; Wasko & Faraj, 2000).

A sense of *we*-ness involves the development of a shared identity (Hatch & Schultz, 1997; Van Maanen & Barley, 1985), which includes a set of shared attributes and values that are meaningful to the group and that help create and preserve a system of meaning that binds people together (Albert, Ashforth, & Dutton, 2000). This is particularly important in social groups that are informally structured around shared interests by individuals whose membership is voluntary (Friedman & McAdam, 1992), because identification with a group can act as an intrinsic motivator for individuals to contribute and not simply have a "free ride" (Kogut & Zander, 1992; Moreland & Levine, 2002).

In virtual contexts though, developing the sense of *we*-ness, which is crucial to support participation, can be difficult (Forman, Ghose, & Wiesenfeld, 2008; Jarvenpaa & Leidner, 1999; Jarvenpaa & Staples, 2000). Meaningful knowledge sharing among professionals often requires situated understanding (Bechky, 2003), a process that is difficult in the asynchronous, low-context setting of online text exchange. Smith (1999) has tracked thousands of Usenet groups and observed, "In many cases a newsgroup is a barren or cacophonous space" (p. 201). He reported that a fifth of groups are entirely empty, and most attract fewer than 50 contributors and produce less than 20 messages per month.

Some of the reasons for the difficulty of creating a sense of *we*-ness in online communities are the lack of traditional social cues, shared artifacts (Stamps & Lipnack, 2005), and shared symbols signaling a common identity (Pratt, 1998). Other developmental obstacles that can hamper information sharing and lead to an online group's quick demise include lack of familiarity among individuals, distinctive thought worlds, disparities in verbal skill, differing cultures, status differences, and challenges associated with physical distance (e.g., Earley & Gibson, 2002; Gruenfeld, Mannix, Williams, & Neale, 1996).

For these and other reasons, a number of researchers have argued that many forums fail to function as "communities" and instead operate as very loose networks with (at best) weak social ties (e.g., Jones, 1997; Jones, Ravid, & Rafaeli, 2004). Some in fact argue that a sense of *we*-ness is impossible to develop and cultivate in online contexts. For example, in the HomeNet project, a multiyear research study at Carnegie Mellon University, Kraut et al. (1998) examined the Internet usage of about 100 families in Pittsburgh during their first few years online and concluded that the Internet had a negative impact on families' and friends' interactions, and that the use of the Internet at home caused a small decline in social and psychological well-being (Kraut et al., 1998). This is very much aligned with Richard Putnam's famous claim in his essay 'Bowling

Alone' (1995) that, partly because of technology, Americans belong to fewer organizations, know their neighbors less, meet with friends less frequently, and even socialize with their families less often.

At the same time, many find these results and arguments both disquieting and at odds with both popular beliefs and personal experiences. Hence, users of the WELL (one of the oldest virtual communities, started in the mid-1980s and still in operation) report quite different experiences from their use of the Internet. As chronicled by Rheingold (1993), members of the WELL offer each other social ties, friendship, and emotional support to the point that the community felt as real as any other locale-based community. Similarly, Galegher et al. (1998) report the intense emotional and practical support offered by the online networks of electronic support groups. And Lam (1998) of the Pacific News Service finds that the Internet is being used to create a global Vietnamese community among many of the 2.5 million Vietnamese who were displaced by the Vietnam War and who are now living on five different continents. Through their use of websites devoted to Vietnamese history, culture, news, and community, Vietnamese immigrants have generated a "Virtual Vietnam," establishing social links and reconnecting with their cultural heritage. In all these virtual communities, members developed social ties and a sense of belonging.

Similarly, researchers found that online groups can develop a strong and meaningful sense of collective identity through interaction (e.g., Postmes, Spears, & Lea, 1999). Wenger, who extensively studied communities of practice, mostly collocated, such as Yucatán midwives, native tailors, navy quartermasters, and meat cutters (Lave & Wenger, 1991), has recently turned to the digital realm. In his *Digital Habitats: Stewarding Technology for Communities* (Wenger, White, & Smith, 2009), he and his coauthors claimed that technology has changed what it means for communities to "be together" and that digital tools are now part of most communities' habitats.

How can we explain these contradictory findings and differences in experiences of the same (Internet) technology? One answer could be, as suggested by Orlikowski and Lacono (2000), that stories of the WELL, electronic support groups, and Virtual Vietnam are descriptions of technologies-in-use while the HomeNet project's measures of "Internet use"—number of hours connected to the Internet—do not tell us what people were actually doing with the Internet and how they were using it—whether they were surfing the Web, shopping for books, interacting with friends, participating in an electronic support group, and so on. This is part of the answer, and we certainly need to look at specific uses enacted by particular people in particular times and places to understand the types of bonds they develop with faraway others. We also argue that another part of the answer is related to the enactment, or not, of the writing mechanisms in online exchanges. As suggested by Carr in *The Shallows* (2010), the Internet creates many opportunities for interruptions, and it does not offer us the best conditions for enacting the writing mechanisms. However, reading the exchanges taking place in communities with a strong sense of *we*-ness such as the WELL, open source software projects, or on OpenIDEO shows that the writing mechanisms are enacted fully. Below, we examine an online forum dedicated to knowledge management to show how the writing mechanisms can support the development of a sense of *we*-ness in virtual contexts.

BUILDING A SENSE OF *WE*-NESS IN AN ONLINE FORUM

KM Forum: From Sharing Information to Becoming a Community

In this chapter, we chose to focus on KM Forum, an online forum of people exchanging information and providing help and expertise about knowledge management. KM Forum belongs to an emerging genre of online forums (Fayard & DeSanctis, 2005) geared toward meeting the development needs of professionals with common interests and complementary knowledge needs (Gray & Tatar, 2004; Herring, 2004). Sometimes referred to as "professional development forums," these are a modern incarnation of electronic mailing list discussion groups, such as those created by Listserv, initiated by academic scholars in the early days of the Internet (e.g., Hert, 1997). The knowledge-driven exchanges sometimes also spur strong emotions, just as we saw with the flame war we examined in Chapter 7 and in some of the letters exchanged in HBC (see Chapter 3). (And as we have seen in other chapters, knowledge communities such as the Republic of Letters or OpenIDEO are also bound by a commitment to similar values and are passionate about their goals.)

Today, these professional development forums are vast, covering innumerable specialty topics for a wide array of professional groups. People participate in these forums voluntarily and intermittently for the purpose of acquiring information, skills, and other resources relevant to their work interests (Lakhani & von Hippel, 2003). KM Forum constitutes a conservative testing ground for our notions about the sense of *we*-ness in online communities: Indeed, if we can show *we*-ness in a forum whose members do not produce anything together, we can conclude that this sense will be even stronger in productive communities such as open source software development or OpenIDEO.

KM Forum[2] was founded in August 2000 by a senior executive in a computer services company and focuses on knowledge management in India. Most participants in the forum are located in India (e.g., Bangalore, Hyderabad, Secunderabad, New Delhi, Calcuta, Jamshedpur, Lucknow), but contributors are also located outside of India (e.g., Bahrain, United States). Anyone with access to the Internet can read and post messages to KM Forum. There is no fee to join. All interactions on the forum take place through writing. Members must go to the forum website to read and post messages, which includes text and limited graphics capability.

Contributors to KM Forum come from a variety of organizations, including local companies, universities, government, consultancies, and large multinationals (e.g., HCL Perot Systems, Ernst & Young, Deutsche Bank AG). Some of the participants have met in person and work in the same company; others seem to know each other by reputation, even if they have not met. Contributors discuss a range of matters related to knowledge management, including KM concepts and principles, technical standards, systems implementation, applications to specific work settings, and career issues. There are many contextual references to knowledge management in India. To protect the anonymity of the contributors, names are substituted with fictitious names.

Our analysis focused on KM Forum's founding period and initial growth, until it reached a steady state of contributions. This period is appropriate for tracing the development of a sense of *we*-ness. In all, we examined 15 months of message content. During this period, there were 527 messages posted by a total of 123 contributors (Fayard & DeSanctis, 2005). Like other professional development forums, participation is characterized by the regular activity of a small core group of members (Gray & Tatar, 2004). Gopal, the founder, is also the most active participant, and moderates as well as facilitates the forum. He also sets the tone with frequent messages praising the growth of the forum. Apart from Gopal, several other active participants seem to share a strong feeling of being part of a community, often referring to it in their messages. The style of the forum is polite and informal. People greet each other, sign their messages, thank people in advance, and give feedback. They often refer to the group as a whole.

Our discourse analysis shows how the members of KM Forum were able to enact a sense of *we*-ness through their communicative written practices. In particular, our analysis shows how through writing, the forum's members developed several practices through which they enacted a sense of *we*-ness: self-referring practices, developing a shared history, and strengthening relationships. As we show, the writing mechanisms were central in the performance of these practices.

Self-Referring

KM Forum contributors share a strong feeling of being part of a community—as a participant puts it, "A group like this serves as solace, sounding board (virtual friend, philosopher, and guide)" (159)[3]—borne out by their heavy use of collective language (e.g. "I am raising some issues that *we* might need to discuss in a serious way" [352]—our emphasis). Shared identity comes about as speakers define themselves in relation to the group, especially through self-referring practices. For example, when a member writes, "Dear KM Forum members," he is presupposing and giving existence to the KM Forum as a group to which other people belong. In that sense, self-referring is purely performative.[4] Indeed, by asserting their identity, members construct the reality of it. There is a similar performative process taking place when a participant writes, "I wish all members a happy new year." This is particularly important in an online context, where language is "doing" in the truest performative sense (Herring, 2004; Kolko, 1995). Performativity here is supported by objectifying: By writing down, one objectifies the existence of the community and makes it visible for all to see.

The self-referring practice is enacted in several ways, mostly by referring to the group as a community and by using an enthusiastic tone. In this setting, shared identity can be found in surface-language features that convey intimacy with others, such as reference to "we," "us," or "our group" (Ashforth & Mael, 1989; Weiner & Mehrabian, 1968); in references to a common, larger community ("our KM professional community"); or in references to a locale, such as one's workplace, homeland, or geographic region, such as the reference to India. Participants frequently refer to the group ("Dear KM Forum members" [189], "the people of this community" [134]) and perform general greetings (e.g., "Hi everyone" [4], or "Hello friends" [508]). They also refer from time to time to the broader knowledge-management community: for example, "What should a KM enthusiast do?" (48). Participants often refer to their Indian context, either through cultural references, such as the celebration of Diwali, an Indian festival (116) or through references to cities, such as Pune, Hyderabad, Mumbai, Delhi, and Bangalore (187). Referring to the Indian context or the KM context is central to the enactment of the addressing mechanism because it allows the forum members to understand each other better and therefore feel closer (Wilson, O'Leary, Metiu, & Jett, 2008). The self-referring practice enacted by participants creates a label, "our group," which is a key unifying symbol. In a context of voluntary participation and of few, if any, unifying symbols, the self-referring practice becomes an important means for developing a sense of *we*-ness.

Apart from referring to the community as a community, the tone used is also conducive to the creation of a sense of *we*-ness. Forum participants often express their commitment and enthusiasm toward the group; for example, "We should believe in ourselves" (46), which is again strongly grounded in the objectifying mechanism of writing. The expression of enthusiasm related to self-referring is important because it increases the attraction to the group and the tendency toward agreement (Fulk, 1993). It also shows the importance of the expression of emotions in the development of a community with a sense of shared identity. Self-referring practices are central in the construction of a sense of *we*-ness and the development of a community among people who never meet face to face. In the forum, it is done only through writing and is deeply dependent on the objectifying and addressing mechanisms.

Building a Shared History

Having a shared history is crucial to the emergence of a sense of *we*-ness among participants in online forums. First, it creates a collective library of knowledge, quotes, and opinions to which members can refer. More specifically, new members can read previous messages (objectifying) to get a sense of the topics discussed. As they go through the messages, they can also learn about the expected behaviors on the forum. Newcomers who do not know the forum's history can be given little summaries or references to old threads. They also need to learn that messages need to be framed clearly and be addressed to the most appropriate audience in terms of knowledge and experience. Thus, writing in its both objectifying and addressing mechanisms is key to the construction of a shared history. In fact, history is intrinsically linked to writing because history is often contrasted with oral histories as performed by oral societies.[5]

Yet, how does one create a sense of belonging in an online environment where the other members of the forum are invisible and where the only shared context is the archived messages? It is a sense of shared history that differentiates groups from each other and endows a given group with a definite identity (White, 1992). To develop a shared history, participants enact a repertoire of discursive practices, such as linking (i.e., they refer explicitly to the content of a previous message in their reply; Baym, 1995); quoting (i.e., copying portions of a previous message in one's reply; Herring, 2001); forwarding messages; and weaving (i.e., summarizing the points discussed in a series of messages; Feenberg, 1989). All these practices are enacted through writing and strongly grounded in the writing mechanisms.

KM Forum contributors often perform linking activities. The contributor, by referring to another message (e.g., "This is in continuation to Gopal's reply to Vibha's query" [16]), shows that her message is connected to the forum's discussions and also shows that she is part of the group and aware of discussions taking place in the forum (e.g., "This has reference to the discussion which happened yesterday and the subject matter raised by Gopal today" [42]). It is because writing objectifies emotions and ideas that participants can refer to them explicitly and specifically. Linking is also, to a certain extent, related to the addressing and specifying mechanisms, as participants provide some background information to their audience (addressing) and try to clarify their thoughts (specifying). In some cases, reflecting also builds upon linking, as contributors detect inconsistencies or raise questions about current issues and previous posts.

Participants also practice quoting—copying portions of a previous message in their reply, a practice grounded on the objectifying mechanism. Quoting creates the impression of adjacency, as it juxtaposes two turns (or portions of turns) in the same message (Herring, 2004). Moreover, participants often forward others' messages—either messages that were posted in another forum that the participant wants to share in KM Forum or replies to a message that were sent only to the poster.

While forwarding could not be performed without objectifying, it is also important in specifying an idea: As people are engaged in a discussion, they forward another message to explain their idea better. In some cases, forwarding also emerges from the reflecting mechanism; for example, "As I was reflecting on your message, I realized that it was connected to the topic discussed in this message."

This forwarding activity, which brings side discussions back to the group, extends the space of the forum and gives a "let's all share what we have" feeling of community. Gopal often weaves messages—that is, summarizes different points made in a discussion and provides a synthesis:

> Hi everyone,
>
> I sense that somewhere the discussions that happened took most people by surprise. So I would like to take a few steps back so that some understandings are clarified (especially for those members who have joined out of interest about KM and have no experience in implementation as a discrete activity) for the whole group (We are 79 now!). [...] Any more thoughts anybody? (29)

Gopal's message is a great example of reflecting, where he shares with the group his thoughts on the current stage of the discussion and invites it to reflect and share, too. The rest of the message where he goes in details is a rich illustration of how the specifying mechanism can be enacted. Last, his first sentence, "I sense that somewhere the discussions that happened took most people by surprise," illustrates reflecting as well as addressing, with his reference to other participants who were surprised by the discussions taking place on the forum. His post is a way to provide them with some clarification. Only a few other members perform weaving (i.e., summarizing the points made in a series of messages). Yet, by referring to previous messages, this weaving activity contributes to the building of a shared history.

In KM Forum, writing mechanisms support the emergence of a shared history, not only because objectifying allows participants to constitute a repository of the messages but also because it provides to newcomers a repertoire of styles and ways of interacting. Reflecting is also important in allowing them to look back and share the realization that they have become a community.

Strengthening Relationships

Developing a sense of *we*-ness involves emotions and the building of relationships. Relationship-building is more difficult in online settings, where people miss informal interactions (e.g., Kraut et al., 1988) and nonverbal cues (Short, Williams, & Christie, 1976; Sproull & Kiesler, 1991). In KM Forum, participants are involved in four main activities of relationship maintenance: thanking people, praising (i.e., saying how interesting and/or useful they found prior messages), inviting to contribute, and "helping" (replying and offering the information requested when someone asks a question). Because they create a positive atmosphere with a lot of support, these activities produce the feeling of being part of a community.

Participants often thank other members for their contribution: "I agree with AG" (180), "Good idea from Gopal" (191), "Shiva, that was a valid point" (153). They also provide a lot of positive feedback; for example, "Some really good insight from Sophia (3410), "I completely agree with Mr Kumar's views" (343). They also often reflect upon others' messages, as illustrated by this message: "I quite agree with keen observation and analysis, as well as candid and quite shocking revelation of Anita…Even I have had similar experience in the past" (48) or "Hi all, this is in continuation to Gopal's response to Durga's query" (102).

When providing information, contributors (often Gopal, the founder, and a few frequent contributors) encourage others to engage in a conversation and to reflect on their post, thus often ending their messages with "Any thoughts?" or "Any comments/ corrections/ additions?" By asking for comments, they create a sense that they are invested in others' opinions, as illustrated in the message below by Gopal:

> Hello friends! I came across this site and found it to be pretty interesting. The Demo and ROI calculations are very very interesting…maybe you might find it good to! Please do post your comments. (323)

Here Gopal is not only inviting others to reflect and share their thoughts, but he is also providing a rationale of why he is sharing this information, addressing in that sense the other members. Similarly, when he writes—"Here is a phenomenal link on storytelling from FastCompany…some would find it quite interesting ciao" (326)—he is enacting addressing by referring to other members' potential interests. Participants always post with an audience in mind, and they in fact are inviting other members to engage in a dialogue with them.

KM Forum members also help each other, providing replies in a timely manner. Consider, for example, a discussion of 14 messages over two days that started with a question posted by a newcomer, Rhodit Gaba. He introduces himself, "Hi all, I have just joined the group. I work as a knowledge manager at XYZ," and asks, "How many companies really implemented KM in India, and who are the major players in KM? Thanks. Rhodit" (272). Another participant (also a newcomer) says that he has a similar question; then a series of replies follow:

> KM has been experimented in various hues by various organizations in India. But in my knowledge there are two home grown initiatives I am aware of […] Infosys […] another is my organization […]. Some other examples are CGEY, Microsoft, PwC, McKinsey, Accenture. (274)

All the replies are very detailed and constitute great illustrations of the specifying mechanism. Interestingly, one of the replies refers to the shared history of the group, relying on the objectifying mechanism which allows record keeping: "To find out how Satyam is doing it, please check the earliest KM Forum messages" (274). Rhodit concludes with a very warm and thankful message: "Hi Pals: Hey [...] This is wonderful [...] I like the quick responses [...] Thanks everyone for the response" (288). Rhodit here expresses positive emotions, which are key to creating bonds within the community. In his thank you message, he is summarizing the responses and reflecting on what he learned—both in terms of content as well as in terms of the community—that is, responsive and helpful members. In this exchange, participants try to help the two new members by providing them with rich and useful replies. They help by responding (which is an action in itself), by providing links to websites and white papers, by sharing their own experiences, by summarizing presentations they went to, or by referring to specific experts. All their responses illustrate the four mechanisms of writing—objectifying, addressing, specifying and, in some cases, reflecting.

Relationship management is important in any group or organization, and crucial in building a sense of belonging and commitment. This becomes even more important when people do not see each other and have no opportunities for informal interactions, relying entirely on written interactions with very few opportunities to exhibit friendly behaviors or support (even simple things such as saying hi in the elevator, holding the door, etc.). The KM Forum threads demonstrate how relationships develop and are maintained in an online setting through writing. In KM Forum, participants develop practices to manage their relationships, thereby creating a friendly and helpful culture, which in turn fosters commitment and a feeling of belonging. These practices—self-referring, building a sense of shared history, and strengthening relationships—were all grounded on the four mechanisms of writing—objectifying, addressing, reflecting, and specifying. Because of the performative nature of writing, through these mechanisms, KM participants not only shared information, they constructed a sense of *we*-ness. In that sense, they are very similar to the members of the Republic of Letters, discussed in Chapter 4. Indeed, members of the Republic of Letters, while not always collocated, could develop, through letters at the time, a community (i.e., a network of people who felt connected by similar values and interests). In the section below, in order to broaden our discussion of how people connect and build communities through writing, we revisit the two other cases discussed in previous chapters: open source software development and OpenIDEO, both examples of communities as well.

CONNECTING THROUGH WRITING

In the beginning of this chapter we argued that what changes nowadays is the expanse of communities and the speed with which bonds can form. Information technology has created truly global communities, such as those for open source software development and OpenIDEO. Indeed, a similar sense of *we*-ness as the one we found enacted by KM Forum members can be found enacted—and through similar practices—in the two communities we discussed in Chapters 7 and 8. In this section, we revisit these two cases and show how their members built a sense of *we*-ness through writing in online contexts. We therefore also illustrate the role of the two other powers of writing, expressing emotions and developing knowledge, for building a community. These two cases are also interesting because open source software development is a well-established movement started in the 1980s and OpenIDEO is a more recent community, launched in August 2010. In this section, we also highlight the role of connectors in online settings. As we saw in Chapter 5, some community members—in the case of the Republic of Letters, Mersenne, and Madame du Châtelet—acted as intermediaries among and conduits of information for other members.

The sense of *we*-ness in the various open source software communities is strong. Open source is both a community of practice and a social movement.[6] It is a community of practice because through a process of legitimate peripheral participation newcomers acquire the skills that transform them into full community members (Lave & Wenger, 1991) who signal bugs, write patches of code, and coordinate with others. It is also a social movement primarily because it poses "collective challenges (to elites, authorities, other groups or cultural codes) by people with common purposes and solidarity in sustained interactions with elites, opponents and authorities" (Tarrow, 1994). This definition of social movements emphasizes the importance of solidarity and shared purpose and distinguishes social movements from simple interest groups. Thus, many of the members of this passionate community are actively engaged in posing collective challenges to elites and are knowledgeable about the main ideological stances of the community. For example, one study reports that out of the 1,540 developers surveyed, only 14% said the ideology was not important (David et al., 2003), thus demonstrating the ideological commitment of many free/open source software developers and their common sense of belongingness. The example of the Linux operating system is by now legendary. In 1991, Linus Torvalds, a Finnish computer science student, wrote the first version of a UNIX kernel for his own use. Instead of securing property rights to his invention, he posted the code on the Internet with a request to other programmers to help upgrade it into a working system (Kogut & Metiu, 2001). The response was overwhelming: In 1998 there were 10,000 developers (McHugh, 1998).

If technology helps with the sheer expanse of communities, then it also helps open up the possibility of outsiders or previously excluded groups to join and get involved. In Chapter 5 we showed that writing-based communities allowed outsiders such as women to participate in the exchange the ideas when official, face-to-face forums were closed to them. A similar phenomenon happens today, when people from all over the world can participate in open source software development, such as Wikipedia, and so on. Thus, as of June 2008 there were 722 Linux user groups in 102 countries. According to a study of developers in eight countries, a wide range of skills improve with community participation: writing re-usable code, being aware of legal issues, and having the ability to deal with criticism (David et al., 2007). Women in various parts of the world also form support groups in order to learn about software development and get involved in open source activities. For example, LinuxChix, the most important women-oriented Linux community, has regional chapters in many countries, developed and less developed.[7] While most of the data come from developed countries, India and Brazil are also active. India, in particular, is involved in all types of activities, offering technical courses, strengthening women's managerial and legal skills, helping newcomers ("newbies"), and listing events of interest to group members. The data also document the tailoring of activities to women's specific needs and interests. For example, several groups post feminist material and encourage discussion of such issues as the status of women in free/open source communities and discrimination in the workplace. Through such activities, and through interacting heavily in writing, women have been able to participate to a community that is predominantly male.[8]

OpenIDEO explicitly aims to invite a diverse and global group of members to engage in sharing and exploring ideas and developing concepts to solve issues related to social innovation. On OpenIDEO, the sense of *we*-ness is also very strong. Practices, such as giving feedback and generating ideas, discussed in Chapter 8, when investigating knowledge development by OpenIDEATORS are also central to the development of this sense of *we*-ness because they rely on highly collaborative interactions. Just like open source software developers, OpenIDEATORS also perform the three practices—self-referring, building a shared history, and managing relationships—enacted by the KM Forum members. This sense of *we*-ness, through an emphasis on collaboration, is strongly supported by the OpenIDEO team, as illustrated by reference to the growth of the community—"There's well over 14,000 of you these days. Let's make some noise!"—or by reflecting on the achievements of the community—"OpenIDEATORS have worked long, hard and in unison to compose their concepts. Join us to help decide which ideas are going to best effect positive change."[9] The sense of *we*-ness is clearly phrased with this message posted by the OpenIDEO team after the announcement of the Webby Award: "Our congrats go out to you all—you're what makes us **Community.**" The announcement of the Webby Award was tweeted by an OpenIDEATOR with this comment: "*Can I say this just once: *we* rock!*"

The role of the OpenIDEO team, in particular the community managers and some very active users, is very similar to the role of Gopal and the few active members in KM Forum. This role is also very similar to the role of intermediaries such as Mersenne and Madame du Châtelet, who were connecting people and cross-pollinating ideas. Individuals such as Mersenne, while not the most creative community members, played central roles in the network; similarly, moderators and forum founders are nowadays key to the functioning of online communities (e.g., Fayard & DeSanctis, 2005; Preece, 2000).

As for the other powers of writing, expressing emotions and developing knowledge, we want to stress that all the writing mechanisms whose roles we illustrated in Part 1, when discussing the development of the Republic of Letters as an intellectual community, can all be enacted in today's online contexts. In this chapter, with the case of the KM Forum, we showed how writing can support the development of a community through the construction of a sense of *we*-ness. Yet, as noted in previous chapters, some of these mechanisms can be hindered or forgotten when people interact with the community in certain contexts or with certain technologies. For example, when community members check posts on their mobile devices while on a subway platform or between meetings, they will probably not reply or will reply with very short messages that have little or no feedback and relationship management.

Take the example of Twitter, which allows participants in a community such as OpenIDEO to follow one another. Twitter messages—tweets—are by definition very short, 140 characters, and thus do not support most of the writing mechanisms. However, if tweets cannot be compared to long, detailed letters or long posts or developed comments, they are still written and can be compared to the billet genre (see the sidebar in Chapter 6). If one thinks of tweets as billets, which were often used by members of the Republic of Letters, we can understand their role not as being central in terms of their content but more in terms of supporting relationships and as a complement to longer letters. Tweets and posts on the Facebook walls of other community members then create a shared context, which helps reinforce the addressing mechanism of writing. As one interviewee noted, Twitter and Facebook support people's social peripheral awareness, while posts, comments, and e-mails are tools for developing the relationships. In a sense, such social media tools support or reinforce our sociomental connections, defined by Chayko (2002) in her book *Connecting: How We Form Social Bonds and Communities in the Internet Age* "as a sense of interpersonal togetherness" and of being "oriented toward and mentally engaged with the other" (p. 39). Chayko (2002) argued that we are able to relate to others because our minds have become socioculturally structured and that writing has liberated us from the "restrictions of orality" (p. 10). "It is in thinking similarly to others in this structured and modeled way that we can mentally connect with them without the absolute necessity of face-to-face copresence" (Chayko, 2002, p. 29). While Chayko's analysis and Wilson et al.'s (2008) notion of perceived proximity are centered on dyadic relationships, our analysis shows that similar mechanisms are at play in the development of a sense of *we*-ness in online communities. Many messages on KM Forum as well as other online communities express a sense of like-mindedness and of perceived proximity.

Social media allow members of a community to share an experience, sometimes nearly simultaneously, and thus to develop a connecting structure and a "perceived proximity" (Wilson et al., 2008), which supports and is reinforced through written interactions via e-mails or online forums. Often, the virtual space constructed by the participants' written communication practices and framed by the visual representation of the interface is perceived by members as a metaphorical space where they can meet, exchange ideas, and bond. Many therefore feel comfortable sharing personal, sometimes quite intimate, stories. Hence, a new OpenIDEATOR, who had shared an inspiration based on a personal story, replies to the greeting and thank you message of one of the community managers—"Hi Jeanne, welcome to OpenIDEO and thanks so much for sharing this deeply personal story"—by referring explicitly to a sense of space: "Thank you for providing a space where such sharing feels welcome and safe."

CHAPTER

10

Beyond the Media: The Power of Writing

> Technologies are not mere exterior aids but also interior transformations of consciousness, and never more than when they affect the word.
>
> —Ong (1982/2002, p. 81)

Our journey in this book has covered a long historical period in which the powers of writing—to express emotions, develop knowledge, and create and sustain communities—have flourished in the context of personal correspondences, distributed organizations, and communities. At the same time, over the same historical period, technologies and practices have changed dramatically. Writing and written practices have always evolved with the different tools and media used in various time periods (e.g., stylus and wax tablet, parchment or paper and quill, typewriter, computer, or smart phone). Currently, we experience two main trends in our communicative practices and use of media. The first is an intensification of writing. Indeed, in spite of claims that we are currently in an era of second orality, current practices indicate an increase in the amount of writing across a variety of media and devices—e-mail, online forums, blogs, and even text messages on cell phones. To these add documents and presentations which have become omnipresent in organizations as well as schools and universities. The second trend consists of the quasisynchronicity of written communication brought about by new media. As we suggested in Chapter 6, these two trends have raised concerns about their effects on people's writing ability as well as on associated abilities, such as reflecting, imagining, and analyzing.

Such concerns have been expressed in a variety of contexts. The potential disappearance or at least reduction of the enactment of writing's mechanisms is what William H. Fitzhugh, the publisher of *The Concord Review,* a journal that showcases high school research in history, deplored in an article in *The New York Times:* "Most kids don't know how to write, don't know any history, and that's a disgrace," Mr. Fitzhugh said. "Writing is the most dumbed-down subject in our schools" (Dillon, 2011). As pointed out by Mr. Fitzhugh, the issue is not about learning and accumulating facts but is rather "about developing a sense of historical context, synthesizing findings into new ideas, and wrestling with how to communicate them clearly—a challenge for many students, now that many schools do not require students to write more than five-paragraph essays" (Dillon, 2011). For similar reasons—developing the ability to articulate and communicate complex arguments—companies now send some of their employees to writing workshops. As we have argued in this book, this is what writing is mostly about: an ability to think—that is, to analyze, understand, and develop a reflection about a feeling or an idea. Indeed, as our research shows, writing, long considered an impoverished mode of communication, is more than a recording technology that is only able to support the transmission of simple or explicit information; it is also an intrinsically creative process that supports the ability to think and articulate emotions and ideas in new ways.

Our mode perspective—reconsidering the role of writing and shifting the focus of analysis from the medium (e.g., a piece of paper, an e-mail, a blog) to the mode of communication (written vs. oral)—allowed us to develop a deep understanding of the core mechanisms underlying the writing process. In the first part of the book, we chose to take a historical perspective, analyzing several sets of historical correspondences in order to avoid the noise that the multiplicity of the new media and their constantly changing features might create. We identified four mechanisms of writing—objectifying, addressing, reflecting, and specifying—and showed how they are at the core of the creative process of writing, regardless of the genre and the media. The historical perspective we took in Part 1 provided us with a set of tools that facilitate a deep understanding of the contemporary phenomena we discuss in Part 2. Our approach illuminates both the similarities between written communicative practices in the past and written communicative practices today as well as the differences due to the use of different media. Despite the proliferation of digital media, the four mechanisms of writing remain essential to emotions, knowledge, and communities.

Our approach and the cases discussed in Part 2 show how the four mechanisms of writing—objectifying, addressing, reflecting, and specifying—provide a constructive way of reconciling conflicting claims and findings about the capabilities of written forms of communication for expressing emotions, developing knowledge, and building communities. Thus, by taking a mode perspective, we are able to explain conflicting results about the possibility of feeling close to faraway others (O'Leary,

Wilson, & Metiu, 2011), or about developing complex creative work using written communication (Cummings, Schlosser, & Arrow, 1995) and show how writing is a creative process. Furthermore, the mechanisms of writing we investigated have the potential to revitalize scholars' conception of the role of writing in expressing emotions, developing knowledge, and building communities. We expect that the mode approach will open up new and productive ways of understanding organizational communication as well as fundamental organizational processes, such as emotional dynamics, knowledge sharing and development, and the formation and sustainment of relationships and common identities across physical distance.

In this chapter, we revisit the contradictory imaginations about writing we highlighted in Chapter 1 and show how writing is intrinsically a creative process, which is crucial for today's organizations, where people often work in distributed teams, where emotions and interpersonal bonds are increasingly important, and where innovation and knowledge development are evermore crucial. Last, we highlight implications of the mode perspective we outlined in this book for understanding communicative practices and organizational process and suggest avenues for future research.

THE PROCESS OF WRITING

Writing is usually perceived as being primarily a trace, a means of recording something that already exists, a thought or an emotion. Often when people refer to writing, they refer to the output of writing—the written word—rather than the creative process involved in the production of the written words. The recorded trace is what is often considered writing's main advantage but also one of the main reasons it has been criticized as being a "dead sign" nearly since its inception. Along with a host of literacy scholars, historians of science, psychologists, and philosophers, we argue that writing is more than a recording tool and that it is in fact a creative process enabling thinking and allowing us to come up with new ideas and to articulate them in novel ways as well as expressing and analyzing emotions. As we discussed in Chapter 2, psychologists such as Vygotsky (1962) and Wolf (2008) have shown that with the invention of writing, the human brain has evolved and developed analytical capabilities[1] that allow us to be creative in developing new ideas and theories as well as become more introspective and express emotions. Wolf explained how the accomplished reader developed specialized brain regions, geared to the rapid deciphering of text, and how this also freed the intellectual faculties of the reader: "The act of putting spoken words and unspoken thoughts into written words releases and, in the process, changes the thoughts themselves" (Wolf, 2008, pp. 65–66). Wolf argued that as humans historically started using language to express their thoughts, the generation of new ideas increased. In that sense, writing can be said to be a creative process. Hence, if some have argued that we think with our hands (Ewenstein & Whyte, 2009; Hoptman, 2002; Kelley & Littman, 2001), we argue that we also think through writing—whether it is writing with a pen on a piece of paper or writing with a keyboard.

In the following sections, we will discuss the complex relationship between written and oral communications. First, in order to better understand some of the perceptions of writing as a trace and the associated imaginations about written and oral communications, we will investigate the perception of space and the imaginations about space and time as they provide relevant and productive similarities with the relationship between oral and written communication. Then, we will discuss how Socrates's dialectic method, despite its association with oral communication, enacts writing's mechanisms and can be described as a perfect illustration of the writing process. Last, we will compare writing with other visual practices and argue that all these practices are supporting a creative process of ideas generation and expression of emotions.

Revisiting the Contradictory Imaginations About Written and Oral Communication in Light of the Opposition Between Space and Time

To put in perspective the current views of writing as an impoverished mode, it is useful to observe the way technology has changed our perceptions of another fundamental notion: that of space. Indeed, this imagination of writing as being frozen and dead, in comparison to oral communication, which is seen as a lively flow, echoes similar reflections on space, which is seen as a container and as dead, and time, which is described as life, as a flow (Massey, 2005). The geographer Doris Massey showed how two assumptions are working together here: what she calls the imagination of representation, illustrated by writing, and the representation of space as a surface, influenced by the geometrical definition of space, where space is understood as a Cartesian coordinate system and where each point is defined by a pair of numerical coordinates. As Massey showed, it is easy to compare the surface of the blank page on which you write to space as a static surface and to imagine writing as the representation that is contained and enclosed in this surface. This geometrical surface can be contrasted with time that passes and is in a continual flow and with oral communication, which is also an ongoing movement. Yet, Massey contested such a dichotomy. Space is not a map, not even a series of maps. It consists of numerous ongoing trajectories—subjective and personal, such as our everyday commute or our interpretation of cities, or more political, such as communities' geographies within a city or a country, continuously constructed in practice (see Lefebvre, 1974/2000). Similarly, writing is not the trace on the piece of paper, not even all the pages in the books or all the books in Borges's imaginary and infinite Babel's library (Borges, 1974) or in the world's largest library, the Library of Congress. Writing is the process of writing down; it is a continuous process in which the writer engages in a dialogue with an audience, and it has always evolved since its inception. Moreover, just as space should not be conceptualized as the negative of time (Massey, 2005), writing should not be conceptualized as the negative of oral communication. Writing and oral communication should be thought of together, as implicated in each other. This does not mean that they are identical or that they should be compared; they should be thought in comparison. It is by taking this position that we have been able to understand the powers of writing and how they are grounded on the mechanisms of writing. As we have argued throughout the book, these mechanisms are at play in all attempts at articulating ideas and emotions.

Socrates's Dialectic Method as an Enactment of the Writing Mechanisms

The imagination of writing as mere recording and empty representation has existed since the beginning of writing. The advent of electronic communication media gave a new boost to criticisms of writing—just like the advent of other communication media have in the past (the printing press or the typewriter, for example). In this section, we revisit Plato's perspective that we discussed in Chapter 1, and suggest that Plato, often seen as a critique of writing, can be rightfully considered as a defender of the mechanisms of writing. Plato, the first philosopher to write, is also known as a fierce critic of writing. More interesting is a second tension at the core of Plato's work: Plato's definition of the dialogue as the main means for finding the truth is intrinsically literate. The word *dialogue* comes from the Greek *dialogos* (conversation, dialogue), which is related to *dialogesthai* ("converse," from *dia* "across" and *legein* "speak"). In Plato's work, dialogue refers to the enactment of the dialectic method—attributed to Socrates—through which two individuals develop a better understanding of a problem and ultimately reach the truth.

Socrates developed his dialectical method in contrast to his contemporaries, the Sophists. In ancient Greek, *sophistès* means "the experts of knowledge" and refers to orators and rhetoric teachers who became famous and powerful in the 5th century BC with the development of democracy in Athens. Indeed, in the newly democratic Athens, writing and reading "beautiful" speeches could help one win a vote among the citizens present in the Agora. Sophists were teaching young Athenians the art of rhetoric (i.e., of making strong and powerful speeches whose main purpose was persuasion, not truth). This is the model of discourse that Socrates and Plato despised and fought in their criticism of writing, as they thought that the young people trained by Sophists focused only on crafting emotionally powerful and thus persuasive texts, without caring very much for the content and the logic of the argument.[2] Socrates considered that the speech used by the Sophists and their students was extremely powerful and could bring about violent passions and wrong decisions, even tyranny (Châtelet, 1989).

In Socrates's view, discourse should be given a new status: One should not only speak beautifully and persuasively but most of all elaborate and articulate a grounded discourse. The method for producing such a discourse is the dialogue, where two passions, two opinions, are confronted. In fact, Sophists' speeches, even though they were written down first, are representative of the oral tradition, based on rhythmical syntax and common turns of phrase that facilitate recitation, persuasion, and memorization. In contrast, the dialogue—as defined by Socrates and Plato—was an argumentative (or dialectic) conversation in which the participants appeal to reason and question assumptions and beliefs to reach agreement on definition and understanding.

The Socratic dialogue reveals the complexity of the relationship between orality and literacy. While Socrates and Plato criticized writing as being the tool used by poets and Sophists to write persuasive speeches that do not build on reasoning but only on beliefs and emotions to convince, the dialectic method that they offered enacts all the mechanisms of writing. Through debate—asking and answering questions—beliefs and opinions are objectified and articulated, and this process, argued Socrates, further stimulates critical thinking. The dialogue implies that specific individuals, with their own sets of beliefs, are engaged in the moment of the dialogue. Thus, addressing is central in the dialogue. Through his questions, Socrates led his interlocutors to be clearer, more specific about their assumptions, and was able to identify the flaws in their reasoning and arguments. Socrates's dialogues aim above all to lead individuals to reflect on what they take for granted, and on the beliefs on which they develop their arguments.

Thus, even though the Socratic dialogue was first practiced in oral conversations and defined by Socrates and Plato in opposition to writing (which they saw as freezing the thinking process and not allowing the questioning required to reach the truth), the main elements of the Socratic method (i.e., a process that implies specifying, addressing, and reflecting, and the action of engaging in the exchange by articulating one's ideas and assumptions) can be used in written communication. As we have shown in this book, these mechanisms have been used in written communication throughout the centuries. In that sense, the Socratic dialectic method is not only a communicative method, but it also illustrates the inner dialogue that we experience as we are trying to find the "right" words or sentences to express emotions or articulate ideas. At the same time, the complex and intricate relationships between speech and writing point out to the mythical nature of the foundational role of speech.

Writing and Other Visual Practices at the Core of the Thinking Process

As we show above, writing is not the only mode that supports the thinking process; particular types of oral exchanges can also support the thinking process. Also, professionals such as engineers, architects, and designers have developed particular visual practices to support their thinking process. For example, Henderson (1999) described the visual culture of engineers and their heavy use of drawings and sketches. In her study, she refers to the "meta-indexicality" quality of drawings which play two main roles. First, as conscription devices, they provide individual thinking tools whereby engineers inscribe their explicit and tacit knowledge. Second, as boundary objects (Bechky, 2003; Carlile, 2002), they serve as enablers of group processes and distributed cognition. What Henderson's study as well as other studies of the role of drawing and visuals for occupations such as architects (e.g., Ewenstein & Whyte, 2009) and designers (e.g., Kelley & Littman, 2001; Stigliani & Ravasi, 2009) point to is the role of drawings as thinking devices (Hoptman, 2002), allowing people to articulate their ideas in a similar way that writing down allows us to articulate our ideas. Drawing (and other forms of visual thinking, including prototyping) as practiced by engineers, architects, and designers allows them to objectify their ideas, specify them, and then reflect upon them. Similar to writing, drawing (and prototyping) is always addressed to a specific audience.

While different occupations and different individuals might rely on a specific practice (e.g., drawing or writing; Gardner, 1983), what is central is the creative process through which both emotions and ideas are articulated. This process resonates with Hoptman's (2002) reflection on the different meanings of drawing. Hoptman (2002) distinguished between drawing as a verb, as an action, and what he calls "projective drawing," or "drawing as a noun" (p. 2). Drawing as a verb highlights the process of making, through which something might be found (Hoptman, 2002, p. 2), while projective drawing suggests the depiction of an idea imagined before the act of drawing. While writing is often understood in a way similar to projective drawing, we have argued in this book that writing should be understood as a verb, an action, a process of making, and that it is this process that enables emotions to be expressed, knowledge to be developed, and communities to be built.

NEW MEDIA: WHAT MIGHT WE LOSE?

Using the mode perspective developed in this book, we wonder what the impact of losing mechanisms such as specifying and reflecting will have on human abilities and the organizational processes that writing has supported. The loss of these mechanisms, which are central to the creative power of writing, seems particularly harmful at a time when constant, quick innovation and creativity are essential both for organizations and for society. In this book, our aim was not to play Cassandra and paint a dark picture of our future with technologies. Yet, as noted by McLuhan (1964/1994), to provide a fair evaluation of a new technology or of change, we need to be aware of what is lost as well as what is gained. While we are not asking to get rid of mobile phones, computers, or the Internet, and we are acknowledging the progress brought by new technologies, without which even our collaboration in writing this book would have been much more difficult, we believe we still need to be aware of what we might lose. In the following sections, we will discuss the impact of new media and some of the major potential losses we might experience by not enacting writing's mechanisms.

Writing has never been a fixed practice. Since its origin, writing has been intrinsically connected with various media that supported its enactment. It was made possible by the purposeful development of the alphabet and other technologies, such as paper, pen, and print. Some technological changes, such as paper, facilitated the development of writing. Similarly, new practices, such as the shift from *scriptura continua* (writing with no separation between words) to the form of writing we know (with separation between words), strengthened writing's power. Indeed, moving away from *scriptura continua* not only made readers more efficient—by allowing them to decipher more easily and faster—it also allowed them to develop silent reading and thus to become more attentive and to start thinking more deeply, developing more complex and challenging arguments and expressing more complex and subtle emotions (Carr, 2010).

Keyboards and word processors are also enablers of the writing process. However, as discussed in previous chapters, some of the features of today's technologies and practices do not always support writing and, in some cases, as we will discuss below, might hinder some of its powers. In particular, we will highlight the impact of new media on our reflective capabilities and our ability to emotionally connect with others.

Being Online and Reflecting

One might argue that writing's power corresponds to the era of the book (Eisenstein, 1980). We have now moved to a new era, the era of the Internet, where we can access an incredible mass of information easily and nearly instantly, where we can communicate with people at a distance almost constantly. However, this constant access to immense amounts of information has some impact on our ability to reflect. Although some claim that in today's world people write more than ever (Baron, 2008), Carr (2010) argues that with the Internet we spend more time reading but that it is a radically different reading experience and that it has implications for the way we think. Indeed, he refers to several studies that highlight the link between the sensory-motor experience of the materiality of the written work and the cognitive processing of the text content: while we read online, we do not devote the same attention to it, and we cannot immerse in the text in a similar way. Many studies of people reading on electronic devices, with hyperlinks, show that the readers have a harder time understanding and remembering than if they were reading in a book (see DeStefano & LeFevre, 2007; Landow & Delany, 2001;

Miall & Dobson, 2001). Moreover, the more hyperlinks (which are interpreted by the brain as potential interruptions), the more difficult it is for the reader to understand and remember (Zhu, 1999).

While these studies focus on reading, the connection with writing is easy to make. As we lose our ability to reflect and take less time to think, it becomes harder to develop a long and complex argument or reflect and articulate difficult and complex emotions. Moreover, the cut-and-paste features, as well as the functions, that we discussed in Chapter 6, tend to prevent writers from taking the time to develop an argument and enact the different mechanisms of writing. As Carr eloquently stated, "The development of a well-rounded mind requires both an ability to find and quickly parse a wide range of information and a capacity for open-ended reflection. There needs to be time for inefficient contemplation, time to operate the machine and time to sit idly in the garden" (2010, p. 168).

Dire predictions about the negative effects of the Internet on reading and writing, and more deeply on our cognitive capabilities, are real and the worries largely founded. Yet, we would like to make a distinction between the Internet as a medium and technologies such as word processing and presentation software (which do not require the Internet) and others that are displayed through the Internet—for example, e-mail and blogs. As we have shown in Chapters 7, 8, and 9, we can still enact the writing mechanisms while using new media. Although it becomes harder to enact them online because of the interruptions created by the Internet, it is still possible, as illustrated by cases such as Open Source, OpenIDEO, and KM Forum.

Social Media and Connecting

The Internet, the "network," and our mobile devices also have negative effects on our ability to feel and express emotions, as Turkle (2011) warned us in her recent book, *Alone Together.* Turkle shows that this ability to always be connected is not necessarily a benefit but that it can also drain us as we try to do everything everywhere and all the time. She argues that while we might be free to work from anywhere, perfect "global souls" (Iyer, 2001) or "neo-nomads" (Abbas, 2011), we are also susceptible to being lonely everywhere. She suggests that these constant connections with people whom we know and don't know via various technologies—text messaging, instant messaging, Facebook, Twitter, and e-mail, to cite a few—might lead to a new solitude.

A perfect example of this twist is illustrated in this essay in the *New York Times,* "I Tweet, Therefore I Am":

> On a recent lazy Saturday morning, my daughter and I lolled on a blanket in our front yard, snacking on apricots, listening to a download of E. B. White reading "The Trumpet of the Swan." Her legs sprawled across mine; the grass tickled our ankles. It was the quintessential summer moment,

> and a year ago, I would have been fully present for it. But instead, a part of my consciousness had split off and was observing the scene from the outside: this was, I realized excitedly, the perfect opportunity for a tweet. (Orenstein, 2010)

Technology might not fulfill its promises of connecting us with distant others, and it might even estrange us from closed ones.

Stories like this abound, and "I tweet, therefore I am" echoes "I share, therefore I am," in which people keep updating their status on Facebook. We all know people who, just like one of Turkle's (2011) interviewees, live their lives on their smart phones. We might also think that a few days, or even only a few hours, with no Internet connection would leave us at a loss. However, our argument is to say that if this happens, it is not our destiny. If this is true, it is because we have been using technology in such a way that we tend to constantly communicate and never sit down and think. The potential for connectivity and connection has led us, argues Turkle (2011), to lack the ability to be alone and to gather ourselves, something that we learned through the exercise of writing and silent reading (Havelock, 1963; Martin, 1994; Ong, 1982/2002). The capacity to read silently has allowed human beings to reflect on ideas and also on emotions. Recent research at the University of Southern California's Brain and Creativity Institute showed that emotions such as empathy and compassion can emerge only if people have enough time and the ability to reflect (Marziali, 2009). It is important to be aware of these limitations and consequences because our constant engagement with connecting technology might lead us to lose a lot of our capacities for thinking creatively and for the kind of solitude that is necessary for empathizing.

Yet, as shown in previous chapters, the new media still afford ways to enact the writing mechanisms, thus allowing us to express emotions and develop knowledge through writing and to build relationships and communities. It is up to us to be aware of the limitations of some of the media, of the powers of writing's mechanisms, and to decide if, when, and how we want to enact them. Thus, companies such as Intel have experimented with imposing hours of quiet time, during which workers cannot use the phone or e-mail but can only think and reflect. The feedback has been positive (Iyer, 2012). The need to pause and reflect has also been recognized by technology developers, who now propose software such as Freedom, an application that can disable an Internet connection for up to eight hours.

IMPLICATIONS FOR STUDIES OF NEW COMMUNICATIVE PRACTICES AND ORGANIZATIONAL PROCESSES

The mode perspective we advance in this book is a useful complement to existing theories of communication media that explain the fostering of idea exchange and the flourishing of relationships in online interactions by referring to the features of the medium, such as archiving capacity (Rice & Love, 1987), rehearsability (Dennis, Fuller, & Valacich, 2008), or editability (Rice, 1994). Without dismissing the impact of new media on the speed and scope of sharing, our perspective draws attention to the fundamental mechanisms of the written mode of communication enacted in computer-mediated exchanges. In this sense, the mode perspective can advance our understanding of some of the main challenges related to the impact of online communication in organizations on the expression of emotions, the development of knowledge, and the building of communities. In this section, we focus on three such avenues for future research that, in our view, hold particular promise in terms of deepening our understanding of fundamental organizational processes: the new communication practices supported by the new media, the creation of new forms of interactional spaces, and the process of knowledge sharing and development.

The Mode Perspective as a Lens for Studying Organizational Communication

Language is much more than a tool used to describe and report on reality; it is the essence of organizations, which are phenomena in and of language (Grant, Keenoy, & Oswick, 1998; Keenoy, Oswick, & Grant, 1997; Orlikowski & Yates, 1994; Schall, 1983; Weick, 1987). Organizational researchers have often studied specific aspects of communication, such as the impact of technology on communication (e.g., Sproull & Kiesler, 1991) or the relationship between communication and organizational characteristics (e.g., Rice & Associates, 1984). More recently, there has been a growing interest in discourse and its consequences for organizational life (Boje, Oswick, & Ford, 2004; Cooren, Kuhn, Cornelissen, & Clark, 2011; Taylor & Van Every, 2011), where language is viewed as central to the organizing process. In particular, as organizations are speech communities (Barley, 1983), and as organizing emerges in the interactive exchanges of their members (Taylor & Van Every, 2000), analyzing their discourse provides rich insights on the organizing and structuring processes. This is particularly true in distributed or virtual contexts. Indeed, while few pure forms of virtual organizations exist today (Dutton, 1999), most organizations present some degree of virtuality (DeSanctis, Staudenmayer, & Wong, 1999; Kraut, Steinfield, Chan, Butler, & Hoag, 1999), and e-mail and instant messaging are pervasive ways of communication, even for collocated people.

While organizational scholars tend to focus on organizational texts—spoken and written—as produced and used by managers and employees (e.g., Cooren, 2004; Keenoy et al., 1997; Orlikowski & Yates, 1994; Reed, 2001; Robichaud, 2001), showing their performativity and how they participate to the organizing process, they focus on the content and ignore the variations implied by the modes of communication. However, as we suggested throughout this book, there is a lot to learn from analyzing how organizing is enacted through writing and, in particular, how writing's mechanisms are involved and support the organizing process. Furthermore, the mechanisms of writing can provide a lens to analyze organizational communication in other contexts than expressing emotions, developing knowledge, and building communities. Future research could, in the vein of the research by O'Mahony and Ferraro (2007), which analyzed the written texts of leaders in the free and open source software community to assess the evolution of different conceptions of leadership over time and use the mechanisms of writing to analyze the development and enactment of leadership or the construction of culture in organizational contexts.

Taking a mode perspective can also help analyze another main trend in current communication practices: the mixing of oral and written modes with the development of media that combine text and speech (e.g., e-mails on wireless devices). This trend has prompted some scholars to envision a return to orality, such that e-mail will be replaced by synchronous and asynchronous video (Bolter, 2001). In light of the mode perspective developed in this book, we suggest that what we witness today is not so much a return to orality but rather a change in the way the mechanisms of writing are enacted. Thus, studies focusing on the relationships between oral and written communication in distributed contexts—and how each support, and in which ways, the expression of emotions, the development of knowledge, and the building of communities—would be extremely fruitful. For example, future research could examine the use of e-mail and other writing practices (e.g., online groups, blogs, instant messaging) in distributed teams and how they mix with oral communication practices such as traveling (in person), phone calls, teleconferences, and videoconferences. It would aim to entangle the role played by each of these practices in the development of ideas and of relationships in both distributed and collocated contexts (the latter are becoming increasingly hybrid, with people e-mailing and messaging colleagues sitting only a few meters away).

Concerns with the potential losses brought about by new media should not deter organizational and communication scholars from examining the positive impact of new media in organizations, especially as it concerns the empowerment it can bring to individuals who are immersed in strong institutional contexts. The anthropologist Broadbent,[3] who has a more positive take than Turkle on the role of new media (such as mobile phones, instant messaging, and social networking), highlights the changes these new media introduced in the work life of many employees. In particular, she highlights power issues related to the attempt by organizations to control the use of these new media. Broadbent argues that new media, instead of broadening our social circles, allow us to deepen our relationships. For example, she reports that when people have 100 individuals on their contact list in instant messaging, they only chat with no more than 5 persons on this list (Anderson, 2009). Similarly, according to a recent study on Facebook, most people have a lot more weak ties than strong ones, even if you consider the most lax definition of a "strong tie" as someone from whom you've received a single message or comment (Bakshy, 2012). New technologies, despite some allegations, are not responsible for people's isolation and lack of engagement in public life (see Putnam, 2000). Modern institutions, Broadbent claimed, are the main reason for this isolation. Fifty years ago, professional life, for most people except senior management, meant isolation. Think, for example, of factory workers or office workers who were completely cut off of their "private sphere" once "at work," unable to check on their children or a sick parent. What new media have provided workers is a "democratized intimacy," argued Broadbent, by allowing people to be in touch with their family and friends (via instant messaging or text messaging), hence "breaking an imposed isolation that institutions are imposing them."[4] Such studies open a fascinating field of research on the relationship between new media, power relationships, and control in organizations.

The New Media and the Construction of Hybrid Interactional Spaces

Taking a mode perspective might also lead us to revisit the notions of virtual teams and virtual organizations. Indeed, as discussed in the first part of this book, current distributed collaborations and communities are not new phenomena and have famous ancestors such as the Hudson's Bay Company and the Republic of Letters. While today's media allow faster interactions and broader reach, what makes contemporary collaboration and communities successful are not these media but the fact that their participants enact the mechanisms of writing.

Yet, new technologies allow a tremendous increase in the reach and the speed of interactions, thus leading to a mixing of the digital and physical environments, a blurring of face-to-face and virtual encounters (Weeks & Fayard, 2011), and the creation of new hybrid spaces in which to interact. On the one hand, we are increasingly interacting online, always on, members of a global village, where distance seems to have been tamed. Yet, we call each other from the airport and pause our tweets long enough only for the plane to take off as we fly to meet each other. This is not only about collocation, as today's interactions are neither totally physical nor totally virtual; for example, many coworkers interact in these new complex spaces which include chats at the coffee machine but also being friends on Facebook, being connected on LinkedIn, Skype-ing each other, reading and commenting on each other's blogs, and following each other on Twitter. These hybrid spaces provide a holding environment for geographically distant people, who oftentimes have not met, to develop most of the time through writing a sense of shared identity

(Fayard & DeSanctis, 2010) and "perceived proximity" (Wilson, O'Leary, Metiu, & Jett, 2008). As recent research reveals, people feel as close to distant coworkers as they do to collocated ones (O'Leary et al., 2011). They feel a nearness that belies the physical distance involved. What research on perceived proximity and the construction of a shared place in online contexts shows is that distance is not only physical but also social and emotional, and that people who are geographically far away might feel closer than people who sit next to each other.

The experience of feeling close to faraway others echoes Turkle's finding that individuals can feel more "present" when online than when interacting face to face (Turkle, 1995) and suggests that developing a sense of shared place with collaborators might matter more than being collocated. While developing such a sense of shared place for distributed collaborators requires some practice, maybe training, and intentionality, it is far from impossible. Writing with new media can support the expression of emotions and the development of relationships between people who are geographically distributed and who sometime have never met. As a proof and an inspiration, one can take the contemporary examples of online communities such as Open Source, OpenIDEO, and KM Forum, discussed in previous chapters.

Of course, part of the feeling of proximity with geographically distant others is due to increased bandwidth and accessibility, which makes people feel they are "always on" (Baron, 2008; Nardi & Whittaker, 2002) and never disconnected from faraway others. These changes, as discussed in previous chapters, are merging and redefining the boundaries between physical and virtual spaces in organizational contexts and in society at large (Fayard, in press; Fayard & Weeks, 2011; Lindtner et al., 2008). Yet, distance relationships and distributed collaborations are not new notions, as we discussed in the first part of this book. For example, as discussed in Chapters 4 and 5, members of the Republic of Letters felt they belonged to a shared intellectual space, constructed through their values and beliefs as well as their collaborations and letters. Similarly, the first Christians felt they belonged to the Church, thanks to the letters shared and read aloud (King & Frost, 2002). In these historical contexts, as well as in today's world, virtual or physical spaces are constantly enacted through practices (Fayard, in press), in particular, writing practices, such as letter or short-note writing, texting on a phone, chatting on Skype or writing long e-mails.

New media, by increasing the speed of interactions and multiplying the genres of interactions (e.g., with e-mail, tweets, blogs) is inviting us to redefine virtual or distributed teams. Moreover, what our research on writing shows is that writing plays a key role in allowing the development of a sense of shared place through the expression of emotions and the development of knowledge. Future research could provide a deeper understanding of the role of writing—often in combination with speech—in building interactional spaces for distributed (and collocated) collaborators.

Implications for Knowledge Development and Innovation

Our perspective on writing's mechanisms offers a complementary perspective to the studies that emphasize the importance of geography and collocation for knowledge sharing and development (Allen, 1977; Olson, Teasley, Covi, & Olson, 2002). It emphasizes that writing and speech are intimately related. For instance, Einstein met with Cartan and with Born, and kept an intense correspondence with both. Similarly, few virtual communities exist purely in an online form. Both the historical examples and the contemporary ones suggest that written (online or through letters) and face-to-face interactions are deeply intertwined and complementary in idea exchanges (Fayard & DeSanctis, 2008; Griffith & Neale, 2001; O'Mahony & Ferraro, 2007). For example, writing provides the ground for effective and fruitful face-to-face interactions. Through agendas and draft documents sent before a meeting, writing allows participants to clarify their thoughts and start building a common understanding. This common understanding is developed after the meeting through the writing of minutes and follow-up documents. These various documents objectify the ideas discussed, and constitute a ground for further articulation and development. Often the output of a meeting is a written document. Therefore the relationship between two modes of communication in the process of knowledge development is quite complex, and suggests that views that favor face-to-face communication need to be much more nuanced. Both the mechanisms of writing and the increase in the amount of written communication suggest that examinations of the knowledge development process need to be more comprehensive and versatile.

Such approaches have the potential to contribute in important ways to the understanding of innovation and collaboration in distributed contexts. If people become better at processing large amounts of data quickly and at multitasking (Carr, 2010; Wolf, 2008), studies show how in parallel our ability to think analytically and creatively is hindered. As the need of innovation is a motto of today's organizations, understanding the risk of losing writing's mechanisms and the associated ability to be contemplative, reflective, and imaginative (Ong, 1982/2002; Wolf, 2008) cannot be ignored. While it is true that we can now very easily and quickly access a broad range of knowledge, thanks to the Internet and search engines such as Google, we should not forget the importance of knowing a topic in-depth, making sense of it, and reflecting on it. This capacity of reflecting and interpreting requires time to daydream and for idle contemplation, which allows us to interpret and make sense.[5] Both types of states of mind are crucial in contemporary organizations in order to allow the emergence of new ideas, the nurturing of inventions, and the development of new innovations. Thus, studies that would distill the particular combinations of communication modes—written and oral—and of media that allow both access to a wealth of information as well as the enactment of the writing's mechanisms, could go a long way toward increasing the effectiveness of collaboration and knowledge development and supporting innovation in informal and formal organizations.

BEYOND THE MEDIA

One cannot help thinking about McLuhan's famous phrase, "The medium is the message," in his book *Understanding Media: The Extensions of Man* (1964/1994), in light of recent debates about new media, such as mobile phones, instant messaging, and social networking and how they do or do not support the expression of emotions, the development of knowledge, and the building of communities. What McLuhan said in this phrase, we believe, is that whenever a new medium came along, people got caught up with the "content" it brought and the technology disappeared behind entertainment, instruction, and conversation. However, McLuhan would probably have agreed with Carr's interpretation of his work: "In the long run a medium's content matters less than the medium itself in influencing how we think and act" (Carr, 2010, p. 3). Yet, in our case, the question remains, how to interpret the interactions between the medium (e.g., the Internet, mobile phones) and the mode of communication, writing?

We argue that while the medium does influence and shape society, and the way we think and we act, as McLuhan showed at his time, and as many scholars have recently shown for contemporary media, we should not forget the modes of communication, writing and speaking, as they allow us to analyze the communicative processes and how they support key practices in organizations—expressing emotions, developing knowledge, and building communities—while not being entangled with the features of the technology. The focus on the mode allows us to better understand the evolution of writing over time and how, despite the media changes, writing has continuously been central to humans' creative thinking process.

While we recognize the *situatedness* of the enactment of writing, we suggest that it is important to be aware of the mechanisms afforded by writing—objectifying, addressing, reflecting, and specifying—whether or not we choose to enact them and how we choose to enact them in different contexts. Writing, as we have shown throughout this book with a variety of examples, case studies, and interviews, is not only a major invention for human beings, it is also a deeply creative process. It has evolved over time, as the tools and media to write have changed. However, objectifying, addressing, reflecting, and specifying, the key mechanisms of writing, have been enacted across periods and media—from the stylus and the wax tablet to the quill and the parchment, the paper and the typewriter to the keyboard and the screen.

Even though some features of the new media might limit or hinder the enactment of the mechanisms—for example, the small screen and keyboard on smart phones and their context of use tend to reduce addressing, specifying, and reflecting—they can still afford to enact the mechanisms. For example, you can decide to take the time to reply thoughtfully and in detail to an e-mail even though you could reply on the spot. You can also choose to write it from your laptop or desktop and draft it several times instead of sending two lines of e-mail. It is important to keep in mind that technology affordances are not obligations but are suggestions for use (Fayard & Weeks, 2007; Gibson, 1986; Norman, 1993), and they can always be ignored or bypassed. The essential question is then, what do we want from the technology and how does it serve our purposes?

It is important to keep in mind that because writing is not an innate ability but a process and practice we acquire and develop with effort (Wolf, 2008), it thus requires attention and continuous investments from individuals and also at the organizational and societal level—as, for example, the Italian merchants in the 14th century who pushed for the development of new teaching methods and a focus on writing—for its powers to come to light (Martin, 1994). Therefore, to maintain the powers of writing, we need to be aware of what technologies and the contexts of use afford (or not) as well as to put some effort in the practice of writing. This is what organizations sending their employees to writing workshops are realizing, as well as schools that reintroduce in their curriculum the importance of writing as a way of thinking creatively.

As we have argued in this book, writing, one of the two fundamental modes of human communication, is crucial in formal and informal organizations, and it possesses four fundamental mechanisms—objectifying, addressing, reflecting, and specifying—which support what we call the three powers of writing: the expression of emotions, the development of knowledge, and the building of communities. We illustrate these four mechanisms and their role in the enactment of the three powers with an analysis of historical correspondences (in Part 1) and contemporary online communities and interviews (in Part 2). In our view, the mode perspective is particularly fruitful in understanding contemporary communicative practices and the changes triggered by new media, as it allows us to go beyond the surface of the constantly evolving new features.

Following the cognitive psychologist Maryanne Wolf (2008), who invites us to reflect on "the value of intellectual skills facilitated through literacy that we don't wish to lose" in order to keep the "profound generativity" (p. 23) supported by writing and reading, we suggest that, by considering the communication media and the repertoire of communicative practices that distributed organizations and informal communities might adopt, both researchers and practitioners need to be attentive to how we can keep enacting the writing mechanisms that are central to expressing emotions, developing knowledge, and building communities.

NOTES

Chapter 1

1. One shouldn't confuse oral traditions with cultures that don't have a written language. Indeed, there are thousands of languages that are only spoken (Ong, 1982/2002), but as people leave their villages more and more, these languages are dying. Of the world's 7,000 languages, 40% are on their way to extinction. This has led two linguists, K. David Harrison and Greg Anderson, to develop a project to document these endangered languages (see http://www.livingtongues.org/hotspots.html). See Ong (1982/2002, pp. 5–15) for a discussion of this distinction.
2. Some of the oldest tablets were found in the city of Ur, and they are now displayed at the British Museum in London, which houses approximately 130,000 written tablets from Mesopotamia. Scholars from around the world come to study the collection. Some tablets kept records for beans, beer (which was the staple drink in Mesopotamia and was issued as rations to workers), and other supplies. Some of the tablets also indicated the beginning of civil service, with some showing how much the state paid some workers, which allowed accountants and civil servants to keep track of public finances.
3. *The First Cities and States: A History of the World*, BBC Program, Radio 4, February 2010, http://www.bbc.co.uk/iplayer/episode/b00qb5y1/A_History_of_the_World_in_100_Objects_The_First_Cities_and_States_(4000_2000_BC)_Early_Writing_Tablet/; see also Martin, 1995, pp. 8–15.
4. *CTIA–The Wireless Association® Announces Semi-Annual Wireless Industry Survey Results* (2010).
5. Study by ARCEP, January 2011.
6. According to Plato (1940b), knowledge is reached by remembering the unchanging Forms or Ideas, or relationships, through a process called dialectics, as illustrated by the allegory of the cave (*Republic* 7.514a).
7. For this 2010 report by the Radicati Group see http://www.radicati.com/.

Chapter 2

1. This is similar to the possibilities that writing affords to literate societies (in contrast to oral societies) to develop formal logics and mathematics (Goody, 1987).
2. Of course written—or oral, for that matter—communication can be used to cultivate opacity (often with very carefully chosen words) instead of clarity. We focus here on the cases in which the writer wants to be understood.

Chapter 3

1. In a letter from the headquarters from 1638, commanders were asked to answer the Company's letter paragraph by paragraph.
2. Unless specified otherwise, all quotes come from the volume 'Letters from Hudson Bay' (1965). We designate letters exchanged by the HBC by the letter's number and by the page number in which it appears in the volume.
3. For a complementary analysis of Virginia Woolf's correspondence and an analysis of Kafka's correspondence, see Fayard and Metiu (2009).
4. Unless specified otherwise, all quotes come from *Congenial Spirits: The Selected Letters of Virginia Woolf* (Trautmann Banks, 1990). We reference to these letters by their date.
5. Dialogue in this context does not refer to the oral nature of a conversation but to the notion of an exchange between two subjectivities discussing a topic.

Chapter 4

1. And not only Western societies—Chinese society is marked by a similar emphasis on written knowledge.
2. Few events in ancient history are as controversial as the destruction of the Library because the historical record is both contradictory and incomplete. Seneca (AD 49), who was the first writer to mention it (and that was nearly 100 years after the alleged event), said that definitely 40,000 books were burned. It is unclear when the fire took place and if it was an accident or intentional. The debate has still not been resolved among historians.
3. Much of subsequent Western philosophy is a reaction to Descartes's writings, hence the reference to him as the "Founder of Modern Philosophy" and the "Father of Modern Mathematics." His influence on mathematics is reflected in the Cartesian coordinate system used in plane geometry and algebra being named after him. He was one of the key figures in what was called by some the "new philosophy," which corresponds to the rejection of the doctrines developed during antiquity and followed until the Middle Ages and the development of new theories, such as the ones developed by Copernicus, Brahe, and Newton, which led to the foundation of modern science.
4. Elisabeth of Bohemia (1618–1680) was the eldest daughter of Frederick V, who was briefly elected King of Bohemia, and Elizabeth Stuart. She is described as a woman of utmost intelligence, speaking six languages and having an aptitude for mathematics and art.
5. We designate letters exchanged by the Hudson's Bay Company by the letter's number and by the page number in which it appears in the volume. All letters from Hudson's Bay Company hereafter quoted in this chapter are excerpted from 'Letters from Hudson Bay' (1965).
6. An indent is an official order for specified goods.
7. Hereafter, all references to letters between Cartan and Einstein in this chapter are excerpts from their correspondence, *Élie Cartan–Albert Einstein, letters on absolute parallelism, 1929–1932*; see Debever (1979).

Chapter 5

1. New Science was at the core of Enlightenment; it involved the scientific study of nature, including man and society. Under the guide of Bacon, Newton, and Locke, learned men and women undertook the replacement of superstition and ignorance with knowledge acquired through the application

of the scientific method—that is, through the systematic application of reason, criticism, skepticism, and the gathering of replicable experiments.

2. The practice is similar to PhD exchange visits today.
3. Universalism stated that knowledge claims are judged (accepted or rejected) through "preestablished impersonal criteria consonant with observations and with previously confirmed knowledge" (Merton, 1942, p. 270). Communism referred to the communitarian aspect of science: "The substantive findings of science are a product of social collaboration and are assigned to the community" (Merton, 1942, p. 273). Organized skepticism argues that scientific judgments should be based on facts. Disinterestedness puts forth the idea that reward comes through recognition and esteem, not through monetary gains.
4. Incidentally, "In Britain, the Secret Office was created in 1653 to open and copy foreign correspondence…Its staff included translators, cryptographers, and even engravers to recreate the seals broken open by government agents" (Heckendorn Cook, 1996, p. 170).
5. Coser (1965) has argued that interaction with peers is required for the development of ideas. He has shown this with studies of various literary and political groupings of the last 300 years.
6. The blurring of the private and the public started with some of the earliest preserved correspondences. For instance, the epistolary works of Pliny the Younger are the first collection of letters pertaining to private affairs in ancient Rome, which the author himself published (de Pretis, 2003). Cicero, too, had planned to publish part of his correspondences, but because he died before he could do so, his secretary was the one who released the papers (de Pretis, 2003).
7. Nicolas-Claude Fabri de Peiresc (1580–1637) was an astronomer whose most famous accomplishment was to produce the first map of the moon that was based on observations with a telescope.
8. Throughout the chapter, translations are our own.
9. Marie Curie (1867–1934) was the first woman professor at the Sorbonne: She took over the chair of her husband, Pierre Curie.
10. The Encyclopedists were a group of French philosophers who collaborated in the 18th century in the production of the Encyclopédie under the direction of Denis Diderot (1713–1784). The Encyclopédie attempted to provide an alphabetical treatment of the whole field of human knowledge from the standpoint of the Enlightenment.
11. In this section, we draw on Bodanis's 2007 authoritative biography of Madame du Châtelet.
12. Hereafter, excerpts are from *Lettres Inédites de Madame du Châtelet* (1818).
13. We use two sources for this analysis. The first is the letters examined by Bonnel (2000), and the second is *Lettres Inédites de Madame du Châtelet* (1818), published by Gallica, the Bibliothèque Numérique Universelle (Universal Digital Library), a project by the National Library of France.
14. Madame du Châtelet to Maupertuis, Letter IV, 1734. *Lettres Inédites de Madame du Châtelet* (1818).
15. Madame du Châtelet to Maupertuis, Letter X, 1734. *Lettres Inédites de Madame du Châtelet* (1818).
16. Madame du Châtelet to Maupertuis, Letter XI.,1734. *Lettres Inédites de Madame du Châtelet* (1818).

17. Charles François de Cisternay Du Fay (1698–1739) was a French scientist and the superintendent of the Jardin du Roi of Paris. He made the important discovery that there are two kinds of electricity, one produced by glass (vitreous) and the other by resin (resinous).
18. Charles-Marie de la Condamine (1701–1774) was a French naturalist and mathematician who accomplished the first scientific exploration of the Amazon River. In 1735, he was part of an expedition sent to Peru to determine the length near the equator of a degree of the meridian.

Chapter 6

1. See http://www.washingtonpost.com/blogs/faster-forward/post/twitter-users-send-200-million-tweets-per-day/2011/07/01/AGEFBUtH_blog.html.
2. See the Radicati Group's report for data from April 2010 (Radicati, 2010).
3. Abbreviation used to refer to the text messages sent by phone through Short Message Service.
4. See http://www.nytimes.com/2011/01/08/education/08research.html?emc=eta1.

Chapter 7

1. We are using the initials of the participants in the discussion.
2. Application Programming Interface.
3. In a letter from December 4, 1926, to Born, Einstein writes that he cannot accept quantum mechanics because of "an inner voice."
4. "He loves me not." Interview with Sophie Calle by Angelique Chrisafis, *The Guardian*, June 15, 2007. Retrieved from http://www.guardian.co.uk/world/2007/jun/16/artnews.art.

Chapter 8

1. Technology can topple tyrants: Jimmy Wales an eternal optimist. (2011, November 7). *Sydney Morning Herald*. Retrieved from http://www.smh.com.au/technology/technology-news/technology-can-topple-tyrants-jimmy-wales-an-eternal-optimist-20111107-1n387.html.
2. The information presented in this section is partly based on the case study OpenIDEO (Lakhani, Fayard, Levina, & Pokrywa, 2012) as well as on our observations of the public posts. The analysis of the communicative practices is based on the exchanges posted on the OpenIDEO platform. The material on OpenIDEO presented in this chapter is based upon work supported by the National Science Foundation under Grant No. 1122413. Any opinions, findings, and conclusions or recommendations expressed in this material are those of the authors and do not necessarily reflect the views of the National Science Foundation.
3. See http://www.ideo.com/expertise/social-innovation/.
4. See http://colinraney.com/2010/08/02/going-open/.
5. Names of OpenIDEATORS in the text are anonymized.
6. See http://www.openideo.com/faq.
7. This phase was introduced in the Bone Marrow Donor challenge.

Chapter 9

1. In the spring of 1915, the French government created and promoted *marraines de guerre* (war godmothers). These war godmothers were supposed

to "adopt" a soldier without a family and write to him regularly, providing personal support.

2. The description of KM Forum and the analysis of the communicative practices of KM Forum members is based on the analysis developed by Fayard and DeSanctis (2010).
3. Numbers in parentheses refer to the number of the message in the forum.
4. It is a similar action as the one described by Austin (1962) with the example of naming a ship. For example, if you say, "I name this ship the *Queen Elizabeth*" (and the circumstances are appropriate in certain ways), you not only utter a sentence but you do something—namely, you perform the act of naming the ship.
5. Historians usually draw the line of separation between prehistory and history with the invention of writing. Moreover, as discussed in Chapter 1, writing emerged from the political expansion and growth of societies in Mesopotamia and Egypt, where the bureaucracy grew and where there was a need to develop a reliable method to transmit information, record financial data, and keep historical accounts.
6. The description of the open source software development communities is based on the analysis developed by Metiu (2010).
7. See http://www.linuxchix.org/regional-chapters.html.
8. For a more detailed analysis of the status and participation strategies of women in the open source software community, see Metiu and Obodaru (2008).
9. See http://www.openideo.com/fieldnotes/openideo-team-notes/openideo-currents-20.

Chapter 10

1. For example, Wolf argued that as the Sumerians and Egyptians developed writing, they literally "crisscrossed" their cortex. Similarly, the psychologist Feggy Ostrosky-Solis and her colleagues show that learning how to read and write "powerfully shape adult neuropsychological systems" (Ostrosky-Solís, Arellano García, & Pérez, 2004, p. 33).
2. Hence the contemporary definition of *sophism* as a clever but false argument, especially one used deliberately to deceive.
3. Stefana Broadbent discussed these issues in a talk at TEDGlobal in 2009. For the video and script of her talk see http://www.ted.com/talks/stefana_broadbent_how_the_internet_enables_intimacy.html.
4. The quotes are excerpts from Stefana Broadbent's talk at TEDGlobal in London in 2009. See note 3.
5. Interestingly, the birth and development of philosophy in Ancient Greece is associated with the importance of leisure in the Greek ideal. In his *Nichomachean Ethics,* Aristotle (1999) suggested that the highest goal in life is leisure, as it gives us the opportunity to live a life of contemplation that includes the pursuit of literature, music, the arts, and science.

REFERENCES

Abbas, Y. (2011). *Le néo-nomadisme*. Paris, France: Fyp éditions.

Ahuja, M. K., & Carley, K. M. (1999). Network structures in virtual organizations. *Organization Science, 10*, 741–757.

Albert, S., Ashforth, B. E., & Dutton, J. E. (2000). Organizational identity and identification: Charting new waters and building new bridges. *Academy of Management Review, 25*, 13–17.

Allen, T. J. (1977). *Managing the flow of technology: Technology transfer and the dissemination of technological information within the R&D organization*. Cambridge, MA: MIT Press.

Altman, J. G. (1982). *Epistolarity: Approaches to a form*. Columbus: Ohio State University Press.

Anderson, B. (1983). *Imagined communities: Reflections on the origin and spread of nationalism*. London: Verso.

Anderson, K. (2009). The democratization of intimacy. *The Digital Content Blog: The Guardian*. Retrieved from http://www.guardian.co.uk/media/pda/2009/jul/23/socialnetworking-skype

ARCEP. (2011, January). Observatoire Trimestriel des marchés des communications électroniques en France: 3ième trimestre—Résultats définitis [Study on the electronic communication markets in France: Final results for the third trimester]. Les Actes de l'ARCEP (Autorité des Communications électroniques et des Postes).

Archak, N. (2010, April 26–30). *Money, glory and cheap talk: Analyzing strategic behavior of contestants in simultaneous crowdsourcing contests on TopCoder.com*. Paper presented at the Proceedings of the 19th International Conference on World Wide Web, Raleigh-Durham, NC.

Aristotle. (1999). *Nichomachean ethics* (2nd ed., T. Irwin, Trans.). Cambridge, MA: Hackett Publishing.

Ashforth, B. E., & Humphrey, R. H. (1995). Emotion in the workplace: A reappraisal. *Human Relations, 48*, 97–125.

Ashforth, B. E., & Mael, F. (1989). Social identity theory and the organization. *Academy of Management Review, 14*, 20–39.

Ashkanasy, N. M., Zerbe, W., & Härtel, C. E. J. (Eds.). (2002). *Managing emotions in the workplace*. Armonk, NY: ME Sharpe.

Auroux, S., Deschamps, J., & Kouloughli, D. (1996). *La philosophie du langage* [The philosophy of language]. Paris: Presses Universitaires de France.

Austin, J. L. (1962). *How to do things with words: The William James Lectures delivered at Harvard University in 1955* (J. O. Urmson, Ed.). Oxford: Clarendon.

Badinter, E. (2006). *Mme du Châtelet, Mme d'Épinay ou l'ambition féminine au XVIIIe siècle* [Mme du Châtelet, Mme d'Épinay or feminine ambition in the 18th century]. Paris: Flammarion.
Bakshy, E. (2012, January 17). Rethinking information diversity in networks. Retrieved from http://www.facebook.com/notes/facebook-data-team/rethinking-information-diversity-in-networks/10150503499618859
Barley, S. R. (1983). Semiotics and the study of occupational and organizational cultures. *Administrative Science Quarterly, 28*(3), 393–413.
Barley, S. R., & Kunda, G. (1992). Design and devotion: Surges of rational and normative ideologies of control in managerial discourse. *Administrative Science Quarterly, 37*(3), 363–399.
Barley, S. R., Meyerson, D. E., & Grodal, S. (2011). E-mail as a source and symbol of stress. *Organization Science, 22*(4), 887–906.
Baron, N. S. (2008). *Always on.* New York: Oxford University Press.
Baym, N. K. (1995). The emergence of community in computer-mediated communication. In S. Jones (Ed.), *Cybersociety* (pp. 35–68). Newbury Park, CA: Sage.
Bazerman, C. (1988). *Shaping written knowledge: The genre and activity of the experimental article in science.* Madison: University of Wisconsin Press.
Bechky, B. A. (2003). Sharing meaning across occupational communities: The transformation of understanding on a production floor. *Organization Science, 14*, 312–330.
Becker, H. (1984). *Art worlds.* Berkeley: University of California Press.
Berkvens-Stevelinck, C. (2005). Introduction. In C. Berkvens-Stevelinck, H. Bots, & J. Haseler (Eds.), *Les grands intermédiaries culturels de la République des Lettres. Études de réseaux de correspondences du XVIème au XVIIIème siècles* [The great cultural intermediaries of the Republic of Letters. Study of the networks of correspondences from the 16th to the 18th centuries] (pp. 9–28). Paris: Honoré Champion.
Berkvens-Stevelinck, C., Bots, H., & Haseler, J. (2005). *Les grands intermédiaires culturels de la République des Lettres. Études de réseaux de correspondances du XVIème au XVIIIème siècle* [The great cultural intermediaries of the Republic of Letters. Study of the networks of correspondences from the 16th to the 18th centuries]. Paris: Honoré Champion.
Beyssade, J. M., & Beyssade, M. (1989). Introduction. In R. Descartes, *Correspondance avec Elisabeth et autres lettres* [Correpondence with Elisabeth and other letters]. Paris: Flammarion.
Bodanis,D. (2001). *E=mc²: A biography of the World's Most Famous Equation.* New York: Berkley Trade.
Bodanis, D. (2007). *Passionate minds: Emilie du Chatelet, Voltaire, and the great love affair of the enlightenment.* New York: Crown.
Boje, D. M., Oswick, C., & Ford, J. D. (2004). Language and organization: The doing of discourse. *Academy of Management Review, 29*, 571–577.
Bolter, J. D. (2001). *Writing space: Computers, hypertext and the remediation of print.* Mahwah, NJ: Erlbaum.
Bonnel, R. (2000). La correspondence scientifique de la marquise du Châtelet: La "lettre-laboratoire" [The scientific correspondence of Marquise du Chatelet: The "laboratory-letter"]. In M-F. Silver & M-L. Girou Swiderski (Eds.), *Femmes en toutes lettres: Les épistolières du XVIIIème siècle* [Women as letter

writers: Female correspondents in the 18th century] (pp. 79–95). Oxford, UK: Voltaire Foundation.

Borges, L. (1974). *Fictions.* Paris: Gallimard.

Born, M., Einstein, A., & Born, H. (1971). *The Born-Einstein letters: Correspondence between Albert Einstein and Max and Hedwig Born from 1916 to 1955 with commentaries by Max Born.* New York: Walker and Company.

Bots, H. (2005). Martin Mersenne, "secrétaire général" de la République des Lettres (1620–1648) [Martin Mersenne, secretary general of the republic of letters (1620–1648)]. In C. Berkvens-Stevelinck, H. Bots, & J. Haseler (Eds.), *Les grands intermédiares culturels de la République des Lettres. Études de réseaux de correspondences du XVIème au XVIIIème siècles* [The great cultural intermediaries of the Republic of Letters. Study of the networks of correspondences from the 16th to the 18th centuries]. Paris: Honoré Champion.

Bots, H., & Waquet, F. (1997). *La république de lettres* [The republic of letters]. Brussels, Belgium: De Boeck.

Bridoux, A. (1953). *Introduction à l'oeuvre et aux lettres de Descartes* [Introduction to Descartes' works and correspondence]. Paris: Gallimard.

Butler, B. (2001). Membership size, communication activity, and sustainability: A resource-based model of online social structures. *Information Systems Research, 12,* 346–362.

Byron, K. (2008). Carrying too heavy a load? The communication and miscommunication of emotion by email. *Academy of Management Review, 33,* 309–327.

Calle, S. (2007). *Prenez-Soin de Vous* [Take care of yourself]. Arles, France: Actes Sud.

Carlile, P. R. (2002). A pragmatic view of knowledge and boundaries: Boundary objects in new product development. *Organization Science, 13*(4), 442–455.

Carlson, N. (2011, January 5). Facebook has more than 600 million users, Goldman tells clients. *Business Insider.* Retrieved from http://articles.businessinsider.com/2011-01-05/tech/30100720_1_user-facebook-pr-goldman-sachs#ixzz1socO1CmS

Carr, N. (2010). *The shallows: What the internet is doing to our brains.* New York: Norton.

Châtelet, F. (1989). *Platon.* Paris: Folio, Gallimard.

Chayko, M. (2002). *Connecting: How we form social bonds and communities in the Internet age.* Albany: State University of New York Press.

Chayko, M. (2008). *Portable communities: The social dynamics of online and mobile connectedness.* Albany: State University of New York Press.

Collins, R. (1998). *A global theory of intellectual change.* Cambridge, MA: Belknap Press.

Constant, D., Sproull, L., & Kiesler, S. (1996). The kindness of strangers: The usefulness of electronic weak ties for technical advice. *Organization Science, 7,* 119–135.

Cooren, F. (2004). Textual agency: How texts do things in organizational settings. *Organization, 11,* 373–393.

Cooren, F., Kuhn, T., Cornelissen, J., & Clark, T. (2011). Communication, organizing and organization: An overview and introduction to the special issue. *Organization Studies, 32,* 1149–1170.

Coser, L. (1965). *Men of ideas: A sociologist's view.* New York: Free Press.

Cramton, C. (2001). The mutual knowledge problem and its consequences for dispersed collaboration. *Organization Science, 12,* 346–371.
Crane, D. (1972). *Invisible colleges: Diffusion of knowledge in scientific communities.* Chicago, IL: University of Chicago Press.
CTIA–The Wireless Association® announces semi-annual wireless industry survey results. (2010, March 23). Retrieved from www.ctia.org/media/press/body.cfm/prid/1936
Cummings, A., Schlosser, A., & Arrow, H. (1995). Developing complex group products: Idea combination in computer-mediated and face-to-face groups. *Computer Supported Cooperative Work, 4,* 229–251.
Curtis, A. (2006). *Virginia Woolf.* London: Haus Publishing Limited.
Daft, R. L., & Lengel, R. H. (1984). Information richness: A new approach to managerial information processing and organization design. In B. Staw & L. L. Cummings (Eds.), *Research in organizational behavior* (pp. 191–233). Greenwich, CT: JAI Press.
Daft, R. L., & Lengel, R. H. (1986). Organizational information requirements, media richness and structural design. *Management Science, 32,* 554–571.
Daft, R. L., Lengel, R. H., & Trevino, L. K. (1987). Message equivocality, media selection, and manager performance: Implications for information systems. *MIS Quarterly, 11,* 355–366.
Daft, R. L., & Wiginton, J. C. (1979). Language and organization. *Academy of Management Review, 4,* 179–191.
Dalton, S. (2003). *Republic of letters: Reconnecting public and private spheres.* Montréal, Quebec, Canada: McGill-Queen's University Press.
Darrow, M. (2000). *French women and the First World War: War stories of the home front (Legacy of the Great War).* Oxford, UK: Berg Publishers.
Daston, L. (1991). The ideal and reality of the republic of letters in the enlightenment. *Science in Context, 4,* 367–386.
David, P. A., Ghosh, R., Glott, R., Gonzales-Barahona, J., Heinz, F., & Shapiro, J. (2007). FLOSS World. Retrieved from http://flossworld.org/deliverables/D31%20-%20Track%201%20International%20Report%20-%20Skills%20Study.pdf
David, P. A., Waterman, A., & Arora, S. (2003). FLOSS-US. Stanford, CA: Stanford Institute for Economic Policy Research, Stanford University. Retrieved from http://www.stanford.edu/group/floss-us/report/FLOSS-US-Report.pdf
Debever, R. (Ed.). (1979). *Élie Cartan–Albert Einstein, letters on absolute parallelism, 1929–1932.* Princeton, NJ: Princeton University Press.
Dennis, A. R., Fuller, R. M., & Valacich, J. S. (2008). Media, tasks, and communication processes: A theory of media synchronicity. *Management Information Systems Quarterly, 32,* 357–380.
De Pretis, A. (2003). "Insincerity," "facts," and "epistolarity": Approaches to Pliny's epistles to Calpurnia. *Arethusa, 36,* 127–146.
Derrida, J. (1987). *The post-card: From Socrates to Freud and beyond.* Chicago, IL: University of Chicago Press.
DeSanctis, G., Staudenmayer, N., & Wong, S. S. (1999). Interdependence in virtual organizations. In C. L. Cooper & D. M. Rousseau (Eds.), *Trends in organizational behavior* (Vol. 6, pp. 81–104). New York: John Wiley.

Descartes, R. (1953a). Méditations: Objections et Réponses [Meditations: Objections and responses]. In A. Bridoux (Ed.), *Oeuvres et lettres* [Works and letters] (pp. 267–547). Paris: Gallimard.

Descartes, R. (1953b). *Oeuvres et lettres* [Works and letters] (A. Bridoux, Ed.). Paris: Gallimard.

Descartes, R. (1953c). Principes de philosophie [Principles of philosophy]. In A. Bridoux (Ed.), *Oeuvres et lettres [*Works and letters] (pp. 553–690). Paris: Gallimard.

DeStefano, D., & LeFevre, J.-A. (2007). Cognitive load in hypertext reading: A review. *Computers in Human Behavior, 23,* 1616–1641.

Devillairs, L. (2001). La Forme méditative de la métaphysique cartésienne [The meditative form of Cartesian metaphysics]. *Philosophiques, 28*(2), pp. 281–301.

Dillon, S. (2011, January 7). Journal showcases dying art of the research paper. *New York Times.* Retrieved from http://www.nytimes.com/2011/01/08/education/08research.html?pagewanted=all

Djelic, M. L., & Quack, S. (2010). *Transnational communities: Shaping global economic governance.* Cambridge, UK: Cambridge University Press.

Dubrovsky, V., Kiesler, S., Sproull, L., & Zubrow, D. (1986). Socialization to computing in college: A look beyond the classroom. In R. S. Feldman (Ed.), *The social psychology of education* (pp. 313–340). Cambridge, UK: Cambridge University Press.

Dumonceaux, P. (1983). Conversion, convertir, étude comparative d'après les lexicographes du XVII siècle [Conversion, to convert, comparative study using the 17the century lexicographers.]. In Centre Méridional de Rencontres sur le XVII siècle, *La Conversion au XVII siècle* (pp. 7–17). Aix en Provence: Publications de l'Université de Provence.

Dutton, W. H. (1999). The virtual organization: Tele-access in business and industry. In G. DeSanctis & J. Fulk (Eds.), *Shaping organizational form: Communication, connection, and community* (pp. 473–495). Newbury Park, CA: Sage.

Earley, P. C., & Gibson, C. B. (2002). *Multinational teams: A new perspective.* Mahwah, NJ: Lawrence Earlbaum and Associates.

Eisenstein, E. L. (1980). *The printing press as an agent of change.* New York: Cambridge University Press.

Ekman, P., Friesen, W. V., & Ancoli, S. (1980). Facial signs of emotional experience. *Journal of Personality and Social Psychology, 39,* 1125–1134.

Elsbach, K. D., Sutton, R. I., & Principe, K. E. (1998). Averting expected controversies through anticipatory impression management: A study of hospital billing. *Organization Science, 9,* 68–86.

Ewenstein, B., & Whyte, J. (2009). Knowledge practices in design: The role of visual representations as epistemic objects. *Organization Studies, 30,* 7–30.

Fayard, A.-L. (in press). A sense of place: The production of virtual and physical spaces. In P. M. Leonardi, B. Nardi, & J. Kallinikos (Eds.), *Materiality and organizing: Social interaction in a technological world.* Ann Arbor: University of Michigan Press.

Fayard, A.-L., & DeSanctis, G. (2005). Evolution of an online forum for knowledge management professionals: A language game analysis. *Journal of Computer-Mediated Communication, 10,* article 2.

Fayard, A.-L., & DeSanctis, G. (2008). Kiosks, clubs and neighborhoods: The language games of online forums. *Journal of the Association for Information Systems, 9*, 677–705.
Fayard, A.-L., & DeSanctis, G. (2010). Enacting language games: The development of a sense of "we-ness" in online forums. *Information Systems Journal, 20*, 383–416.
Fayard, A.-L., & Metiu, A. (2009). Expressing emotions and building relationships over distance: Fixedness and fictionalization in correspondence. In K. Elsbach & B. Bechky (Eds.), *Qualitative organizational research: Best papers from the Davis Conference* (pp. 149–180). Charlotte, NC: Information Age.
Fayard, A.-L., & Weeks, J. R. (2007). Photocopiers and water-coolers: The affordances of informal interactions. *Organization Studies, 28*, 605–634.
Fayard, A.-L., & Weeks, J. (2011, July–August). Who moved my cube? Creating workspaces that actually foster collaboration. *Harvard Business Review*, 102–110.
Feenberg, A. (1989). *The written world: On the theory and practice of computer conferencing.* Oxford, UK: Pergamon Press.
Fineman, S. (1993). *Organizations as emotional arenas.* Thousand Oaks, CA: Sage.
Forman, C., Ghose, A., & Wiesenfeld, B. (2008). Examining the relationship between reviews and sales: The role of reviewer identity disclosure in electronic markets. *Information Systems Research, 19*, 291–313.
Forman, P. (1980). Élie Cartan, Albert Einstein, letters on absolute parallelism 1929–1932. *Bulletin of the Atomic Scientists, 36*, 51.
Friedman, D., & McAdam, D. (1992). Networks, choices and the life of a social movement. In A. D. Morris & C. M. Mueller (Eds.), *Frontiers in movement theory* (pp. 156–173). New Haven, CT: Yale University Press.
Fulk, J. (1993). Social construction of communication technology. *Academy Management Journal, 36*, 921–950.
Galegher, J., Sproull, L., & Kiesler, S. (1998). Legitimacy, authority, and community in electronic support groups. *Written Communication, 15*, 493–530.
Gardiner, L. (1984). Women in science. In S. I. Spencer (Ed.), *French women and the age of enlightenment* (pp. 181–193). Bloomington: Indiana University Press.
Gardner, H. (1983). *Frames of mind.* New York: Basic Books.
Gardner, H. (2008, February 17). The end of literacy? Don't stop reading. *Washington Post*, p. B01.
Gibson, J. J. (1986). *The ecological approach to visual perception.* Hillsdale, NJ: Erlbaum.
Giddens, A. (1984). *The constitution of society: Outline of the theory of structuration.* Berkeley: University of California Press.
Glaser, B., & Strauss, A. (1967). *The discovery of grounded theory: Strategies for qualitative research.* Chicago, IL: Aldine.
Goodman, D. (1989). *Criticism in action: Enlightenment experiments in political writing.* Ithaca, NY: Cornell University Press.
Goodman, D. (1994). *The republic of letters.* Ithaca, NY: Cornell University Press.
Goody, J. (1977). *The domestication of the savage mind.* London: Cambridge University Press.
Goody, J. (1987). *The interface between the written and the oral.* London: Cambridge University Press.

Goody, J., & Watt, I. (1963). The consequences of literacy. *Comparative Studies in Society and History, 5,* 304–345.

Grant, D., Keenoy, T., & Oswick, C. (1998). *Discourse and organization.* London: Sage.

Grassi, M.-C. (2005). *Lire l'épistolaire* [Reading correspondence]. Paris: Armand Colin, 2005.

Gray, J., & Tatar, D. (2004). Sociocultural analysis of online professional development: A case study of personal, interpersonal, community, and technical aspects. In S. A. Barab, R. Kling, & J. H. Gray (Eds.), *Designing for virtual communities in the service of learning* (pp. 404–436). New York: Cambridge University Press.

Griffith, T. L., & Neale, M. A. (2001). Information processing in traditional, hybrid, and virtual teams: From nascent knowledge to transactive memory. In B. Staw & R. Sutton (Eds.), *Research in organizational behavior* (Vol. 23, pp. 379–421). Stamford, CT: JAI Press.

Gruenfeld, D. H., Mannix, E. A., Williams, K. Y., & Neale, M. A. (1996). Group composition and decision making: How member familiarity and information distribution affect process and performance. *Organizational Behavior and Human Decision Processes, 67*(1), 1–15.

Guillen, M. F. (1997). Scientific management's lost aesthetic: Architecture, organization, and the taylorized beauty of the mechanical. *Administrative Science Quarterly, 42,* 682–715.

Habermas, J. (1989). *The structural transformation of the public sphere: An inquiry into a category of bourgeois society.* Cambridge, MA: MIT Press.

Hatch, M. J., & Schultz, M. (1997). Relations between organizational culture, identity and image. *European Journal of Marketing, 31,* 356–365.

Havelock, E. A. (1963). *Preface to Plato.* Cambridge, MA: Harvard University Press.

Haythornthwaite, C., & Kendall, L. (2010). Internet and community. *American Behavioral Scientist, 53,* 1083–1094.

Heckendorn Cook, E. (1996). *Epistolary bodies: Gender and genre in the eighteenth-century Republic of Letters.* Stanford, CA: Stanford University Press.

Hedstrom, P., & Swedberg, R. (1998). *Social mechanisms: An analytical approach to social theory.* Cambridge, UK: Cambridge University Press.

Henderson, K. (1999). *On line and on paper: Visual representations, visual culture, and computer graphics in design engineering.* Cambridge, MA: MIT Press.

Herring, S. C. (2001). Computer-mediated discourse. In D.T.D. Schiffrin & H. Hamilton (Eds.), *The handbook of discourse analysis* (pp. 612–634). Oxford, UK: Blackwell.

Herring, S. C. (2004). Online communication: Through the lens of discourse. In M. Consalvo, N. Baym, J. Hunsinger, K. B. Jensen, J. Logie, M. Murero, & L. R. Shade (Eds.), *Internet research annual* (Vol. 1, pp. 65–76). New York: Peter Lang.

Hert, P. (1997). Social dynamics of an on-line scholarly debate. *The Information Society, 13*(4), 329–360.

Hiemstra, G. (1982). Teleconferencing, concern for face and organizational culture. In M. Burgoon (Ed.), *Communication yearbook* (Vol. 6, pp. 874–903). Beverly Hills, CA: Sage.

Hinds, P., & Mortensen, M. (2005). Understanding conflict in geographically distributed teams: An empirical investigation. *Organization Science, 16,* 290–307.

Hoptman, L. (2002). Introduction: Drawing is a noun. In *Drawing now: Eight propositions* (pp. 11–12). New York: Museum of Modern Art.

Hudson Beatie, J., & Buss, H. M. (2003). *Undelivered letters to Hudson's Bay Company men on the northwest coast of America, 1830–1852*. Vancouver, British Columbia, Canada: University of British Columbia Press.

Hutchins, E. (1991). Organizing work by adaptation. *Organization Science, 2,* 14–39.

Hutton, S. (2004). Emilie du Châtelet's "Institutions de Physique" as a document in the history of French Newtonianism. *Studies in History and Philosophy of Science Part A, 35,* 515–531.

Iyer, P. (2001). *The global soul: Jet lag, shopping malls, and the search for home.* New York: Vintage.

Iyer, P. (2012, December 29). The joy of quiet. *New York Times.* Retrieved from http://www.nytimes.com/2012/01/01/opinion/sunday/the-joy-of-quiet.html?pagewanted=all

Jarvenpaa, S. L., & Leidner, D. E. (1999). Communication and trust in global virtual teams. *Organization Science, 10,* 791–815.

Jarvenpaa, S. L., & Staples, D. (2000). The use of collaborative electronic media for information sharing: An exploratory study of determinants. *Journal of Strategic Information Systems,* 9(2/3), 129–154.

Johansen, R., DeGrasse, R., & Wilson, T. (1978). *Group communication through computers, Vol. 5: Effects on working patterns.* Menlo Park, CA: Institute for the Future.

Jones, Q. (1997). Virtual-communities, virtual-settlements and cyberarchaeology: A theoretical outline. *Journal of Computer Mediated Communication, 3*(3). Retrieved from http://www.ascusc.org/jcmc/vol3/issue3/jones.html

Jones, Q., Ravid, G., & Rafaeli, S. (2004). Information overload and the message dynamics of online interaction spaces: A theoretical model and empirical exploration. *Information Systems Research, 15*(2), 194–210.

Kadushin, C. (1966). The friends and supporters of psychotherapy: On social circles in urban life. *American Sociological Review, 31,* 786–802.

Kadushin, C. (1968). Power influence and social circles: A new methodology for studying opinion-makers. *American Sociological Review, 33,* 685–699.

Kaplan, S. (2011). Strategy and PowerPoint: The epistemic culture and machinery of strategy making. *Organization Science, 22,* 320–346.

Karp, G., Stone, P., & Yoels, W. C. (1977). *Being urban: A social psychological view of city life.* Lexington, MA: D. C. Heath.

Keenoy, T., Oswick, C., & Grant, D. (1997). Organizational discourses: Texts and context. *Organization Science, 4,* 147–157.

Kelley, T., & Littman, J. (2001). *The art of innovation: Lessons in creativity from IDEO, America's leading design firm.* New York: Currency/Doubleday.

Kidd, A. (1994). The marks are on the knowledge workers. *Proceedings of the CHI Conference on Human Factors in Computing Systems* (pp. 186–191). Boston, MA: ACM.

King, J. L., & Frost, R. L. (2002). Managing distance over time: The evolution of technologies of dis/ambiguation. In P. Hinds & S. Kiesler (Eds.), *Distributed work: New research on working across distance using technology* (pp. 3–27). Cambridge, MA: MIT Press.

Klein, M. J., Kox, A. J., & Schulmann, R. (1993). *The collected papers of Albert Einstein. The Swiss years: Correspondence, 1902–1914* (Vol. 5). Princeton, NJ: Princeton University Press.

Kogut, B., & Metiu, A. (2001). Open-source software development and distributed innovation. *Oxford Review of Economic Policy, 17*, 248–264.

Kogut, B., & Zander, U. (1992). Knowledge of the firm, combinative capabilities, and the replication of technology. *Organization Science, 3*, 383–397.

Kolko, B. (1995). Building a world with words: The narrative reality of virtual communities. *Works and Days, 13*, 105–126.

Kraut, R. E., Galegher, J., & C. Egido. (1988). Relationships and tasks in scientific research collaborations. *Human-Computer Interaction, 3*, 31–58.

Kraut, R., Kiesler, S., Mukhopadhyay, T., Scherlis, W., & Patterson, M. (1998). Social impact of the Internet: What does it mean? *Communications of the ACM, 41*, 21–22.

Kraut, R., Steinfield, C., Chan, A. P., Butler, B., & Hoag, A. (1999). Coordination and virtualization: The role of electronic networks and personal relationships. *Organization Science, 10*, 722–740.

Lakhani, K. R., Fayard, A.-L., Levina, N., & Pokrywa, S. "OpenIDEO." Harvard Business School Case 612–066.

Lakhani, K. R., & Jeppesen, L. B. (2007, May). R&D—Getting unusual suspects to solve R&D puzzles. *Harvard Business Review, 85*(5), 30.

Lakhani, K. R., & von Hippel, E. (2003). How open source software works: "Free" user-to-user assistance. *Research Policy, 32*, 923–943.

Lam, A. (1998, November 20). Virtual Vietnam. *All things considered* [Radio broadcast]. Washington, DC: National Public Radio.

Landow, G., & Delany. (2001). *Hypertext, hypermedia and literacy studies: The state of the art, in multimedia: From Wagner to virtual reality.* New York: Norton.

Lave, J., & Wenger, E. (1991). *Situated learning: Legitimate peripheral participation.* Cambridge, UK: Cambridge University Press.

Lea, M., O'Shea, T., Fung, P., & Spears, R. (1992). *"Flaming" in computer-mediated communication.* London: Harvester-Wheatsheaf.

Lefebvre, H. (2000). *La production de l'espace* [The production of space] (4th ed.). Paris: Anthropos. (Original work published in 1974)

Le Naour, J-Y. (2008, March). The wartime godmothers: The soldiers' other family, Les Chemins de la Mémoire, Review. *MINDEF/SGA/DMPA, 181.* Retrieved from http://www.cheminsdememoire.gouv.fr/page/affichepage.php?idLang=en&idPage=13506

Leonard-Barton, D. (1995). *Wellsprings of knowledge: Building and sustaining the sources of innovation.* Boston, MA: Harvard Business School Press.

Leonardi, P. M., & Barley, S. R. (2008). Materiality and change: Challenges to building better theory about technology and organizing. *Information and Organization, 18*, 159–176.

Lepore, S. J., & Smyth, J. M. (2002). *The writing cure: How expressive writing promotes health and emotional well-being.* Washington, DC: American Psychological Association.

Letters from Hudson Bay 1703–1740. (1965). In K. G. Davies & A. M. Johnson (Eds.), *Hudson's Bay Record Society: Vol. 25.* Manitoba, Canada: The Hudson's Bay Company Archives.

Letters of Abelard and Heloise. (1974). (B. Radice, Trans.). New York: Penguin.

Letters of a Portuguese Nun. (2007). (M. Alcoforado, Ed., & E. Prestage, Trans.). Whitefish, MT: Kessinger Publishing. (Original work published 1669)

Lettres inédites de Madame du Châtelet et supplément à la correspondance de Voltaire avec le roi de Prusse et avec différentes personnes célèbres [Unpublished letters of Madame du Châtelet in addition to the correspondence of Voltaire with the King of Prussia and other famous people] (1818). Paris: Imprimerie de Lefebvre.

Levina, N. (2005). Collaborating on multi-party information systems development projects: A collective reflection-in-action view. *Information Systems Research, 16,* 109–130.

Levina, N., & Orlikowski, W. (2009). Understanding shifting power relations within and across organizations: A critical genre analysis. *Academy of Management Journal, 52,* 672–703.

Lévi-Strauss, C (1962). *La pensée sauvage* [The savage mind] (Vol. 2). Paris: Plon.

Lindtner, N., Nardi, B., Wang, Y., Mainwaring, S., Jing, H., & Liang, W. (2008, November 8–12). *A hybrid cultural ecology: World of Warcraft in China.* Paper presented at the CSCW 2008, ACM, San Diego, CA.

Locker, K. (1991, March). *"Sharply and nipplingly" vitupterative letters to subordinates and peers in the correspondence of the British East India Company, 1600–1800.* Paper presented at the CCCC Conference (Conference on College Composition and Communication), Boston, MA.

Lux, D. S., & Cook, H. J. (1998). Closed circles or open networks? Communicating at a distance during the scientific revolution. *History of Science, 36,* 179–211.

Maitlis, S., & Ozcelik, H. (2004). Toxic decision processes: A study of emotion and organizational decision making. *Organization Science, 15,* 375–393.

Martin, H-J. (1994). *The history and power of writing.* Chicago, IL: University of Chicago Press.

Marziali, C. (2009, April 14). *Nobler instincts take time.* Retrieved from http://dornsife.usc.edu/news/stories/547/nobler-instincts-take-time/

Massey, D. (2005). *For space.* London: Sage.

Mazmanian, M., Orlikowski, W., & Yates, J. (2006, July). *Crackberrys: Exploring the social implications of ubiquitous wireless email devices.* Paper presented at the EGOS Conference, Bergen, Norway.

McCorduck, P. (1979). *Machines who think.* New York: Freeman.

McGrath, J. E., & Hollingshead, A. B. (1994). *Groups interacting with technology: Ideas, evidence, issues, and an agenda.* Thousand Oaks, CA: Sage.

McHugh, J. (1998, August 10). For the love of hacking. *Forbes.* Retrieved from http://www.forbes.com/forbes/1998/0810/6203094a.html

McLuhan, M. (1994). *Understanding media: The extensions of man.* Cambridge, MA: MIT Press. (Original work published 1964)

McMillan, D. W., & Chavis, D. M. (1986). Sense of community: A definition and theory. *American Journal of Community Psychology, 14,* 6–23.

Merton, R. (1942). *The sociology of science: Theoretical and empirical investigations.* Chicago, IL: University of Chicago Press.

Merton, R. (1968). *Social theory and social structure.* New York: Free Press.

Metiu, A. (2006). Owning the code: Status closure in distributed groups. *Organization Science, 17,* 418–435.

Metiu, A. (2010). Gift-giving, transnational communities, and skill building in developing countries: The case of free/open source software. In M-L. Djelic

& S. Quack (Eds.), *Transnational communities and the regulation of business* (pp. 199–225). Cambridge, UK: Cambridge University Press.

Metiu, A., & Obodaru, O. (2008). *Women's professional identity formation in the free/open source software community* (ESSEC Working Paper 08009). CERESSEC, Cergy, France.

Miall, D. S., & Dobson, T. (2001). Reading hypertext and the experience of literature. *Journal of Digital Information, 2*(1). Retrieved from http://journals.tdl.org/jodi/article/view/jodi-36/37

Montesquieu, C. (1973). *Lettres persanes* [Persian letters]. Paris: Gallimard, Folio. (Original work published 1754)

Moon, J. Y., & Sproull, L. (2002). Essence of distributed work: The case of the Linux kernel. In P. Hinds & S. Kiesler (Eds.), *Distributed work* (pp. 381–404). Cambridge, MA: MIT Press.

Moorman, C., & Miner, A. S. (1998). Organizational improvisation and organizational memory. *Academy of Management Review, 23*, 698–723.

Moreland, R. L., & Levine, J. M. (2002). Socialization and trust in workgroups. *Group Processes and Intergroup Relations, 5*, 185–201.

Nardi, B., Schiano, D., Grumbrecht, M., & Swatz, L. (2004). Why we blog. *Communications of the ACM, 47*, 41–46.

Nardi, B., & Whittaker, S. (2002). The place of face to face communication in distributed work. In P. Hinds & S. Kiesler (Eds.), *Distributed work: New research on working across distance using technology* (pp. 83–112). Cambridge, MA: MIT Press.

Nie, N., Hillygus, D. S., & Erbring, L. (2003). Internet use, interpersonal relations and sociability: A time diary study. In B. Wellman & C. Haythornthwaite (Eds.), *The Internet in everyday life* (pp. 215–243). Oxford, UK: Blackwell.

Norman, D. A. (1993). *Things that make us smart: Defending human attributes in the age of the machine.* Boston, MA: Addison-Wesley.

Okhuysen, G. A., & Bechky, B. (2009). Coordination in organizations: An integrative perspective. *Academy of Management Annals, 3*(1), 463–502.

O'Leary, M., Orlikowski, W., & Yates, J. (2002). Distributed work over the centuries: Trust and control in the Hudson's Bay Company, 1670–1826. In P. Hinds & S. Kiesler (Eds.), *Distributed work: New research on working across distance using technology* (pp. 25–54). Cambridge, MA: MIT Press.

O'Leary, M., Wilson, J., & Metiu, A. (2011). *Beyond being there: The symbolic role of communication and identification in the emergence of perceived proximity in geographically dispersed work* (ESSEC Working Paper 1112). CERESSEC, Cergy, France.

Oliveira, J. G., & Barabási, A. L. (2005). Darwin and Einstein correspondence patterns. *Nature, 437*(7063), 1251.

Olson, D. (1977). From utterances to text: The bias of language in speech and writing. *Harvard Education Review, 47*, 257–281.

Olson, J., Teasley, S., Covi, L., & Olson, G. (2002). The (currently) unique advantages of collocated work. In P. Hinds & S. Kiesler (Eds.), *Distributed work: New research on working across distance using technology* (pp. 113–135). Cambridge, MA: MIT Press.

O'Mahony, S., & Ferraro, F. (2007). The emergence of governance in an open source community. *Academy of Management Journal, 50*, 1079–1106.

Ong, W. J. (2002). *Orality and literacy* (2nd ed.). New York: Routledge. (Original work published 1982)

Orenstein, P. (2010, July 30). I Tweet, therefore I am. *New York Times.* Retrieved from http://www.nytimes.com/2010/08/01/magazine/01wwln-lede-t.html
Orlikowski, W. J. (2002). Knowing in practice: Enacting a collective capability in distributed organizing. *Organization Science, 13,* 249–273.
Orlikowski, W. J. and Iacono, C. S. (2000). The Truth is Not Out There: An Enacted View of the Digital Economy, in B. Kahin and E. Brynjolfsson (Eds.) *Understanding the digital economy: Data, tools, and research* (pp. 352–380). Cambridge, MA: MIT Press.
Orlikowski, W. J., & Scott, S. V. (2008). Sociomateriality: Challenging the separation of technology, work and organization. *Annals of the Academy of Management, 2,* 433–474.
Orlikowski, W. J., & Yates, J. (1994). Genre repertoire: The structuring of communicative practices in organizations. *Administrative Science Quarterly, 39,* 541–574.
Orr, J. E. (1996). *Talking about machines: An ethnography of a modern job* [Collection on technology and work]. Ithaca, NY: ILR Press.
Ostrosky-Solís, F., Arellano García, M., & Pérez, M. (2004). Can learning to read and write change the brain organization? An electrophysiological study. *International Journal of Psychology, 39*(1), 27–35.
Pagès, A. (1982). La communication circulaire. In J.-L. Bonnat & M. Bossis (Eds.), *Ecrire, publier, lire. Les correspondances (Problématique et économie d'un "genre littéraire")* [Writing, publishing, reading: Correspondences (problems and logic of a "literary genre")] (pp. 344–353). Nantes, France: Université de Nantes.
Pennebaker, J. W. (1990). *Opening up: The healing power of confiding in others.* New York: Morrow.
Pentland, B. T. (1992). Organizing moves in software support hot lines. *Administrative Science Quarterly, 37,* 527–548.
Petrovich, V. C. (1999). Women and the Paris Academy of Sciences. *Eighteenth-Century Studies, 32,* 383–390.
Phillips, N., & Hardy, C. (2002). *Discourse analysis: Investigating processes of social construction.* Thousand Oaks, CA: Sage.
Pinch, T. (2008). Technology and institutions: Living in a material world. *Theory and Society, 37,* 461–483.
Plato. (1940a). The Phaedrus, 227a–275b. In L. Robin (Ed.), *Oeuvres complètes* [Complete works] (Vol. 2, pp. 227–279). Paris: Bibliothèque de la Pléiade, Gallimard.
Plato. (1940b). The Republic, I327a–X621d. In L. Robin (Ed.), *Oeuvres complètes* [Complete works] (Vol. 1, pp. 851–1241). Paris: Bibliothèque de la Pléiade, Gallimard.
Poisson, C. (2002). *Sartre et Beauvoir: Du je au nous* [Sartre and Beauvoir: From I to we]. Amsterdam, The Netherlands: Rodopi Press.
Pondy, L. R., & Mitroff, I. I. (1979). Beyond open systems models of organizations. In B. M. Straw (Ed.), *Research in organizational behavior* (pp. 3–39). Greenwich, CT: JAI Press.
Postmes, T., Spears, R., & Lea, M. (1999). Social identity, group norms, and "deindividuation": Lessons from computer-mediated communication for social influence in the group. In N. Ellemers, R. Spears, & B. Doosje (Eds.), *Social identity: Context, commitment, content* (pp. 164–183). Oxford, UK: Blackwell.

Powell, W. W., & Snellman, K. (2004). The knowledge economy. *Annual Review of Sociology, 30,* 199–220.
Pratt, M. G. (1998). To be or not to be: Central questions in organizational identification. In P. Godfrey (Ed.), *Identity in organizations: Developing theory through conversations* (pp. 171–207). Thousand Oaks, CA: Sage.
Preece, J. (2000). *Online communities.* New York: John Wiley & Sons.
Putnam, R. D. (1995). Bowling alone: America's declining social capital. *Journal of Democracy, 6*(1), 65–78.
Putnam, R. D. (2000). *Bowling alone: The collapse and revival of American community.* New York: Simon & Schuster.
Radicati, S. (Ed.). (2010). *Email statistics report, 2010.* Palo Alto, CA: The Radicati Group.
Reed, M. (2001, August). *The explanatory limits of discourse analysis in organisation analysis: A review.* Paper presented at the Academy of Management, Washington, DC.
Rheingold, H. (1993). *The virtual community.* Reading, MA: Addison-Wesley.
Rice, R. E. (1994). Relating electronic mail use and network structure to R&D work and networks and performance. *Journal of Management Information Systems, 11*(1), 9–20.
Rice, R. E., & Associates. (Eds.). (1984). *The new media: Communication, research and technology.* Beverly Hills, CA: Sage.
Rice, R., & Love, G. (1987). Electronic emotion: Socioemotional content in a computer-mediated communication network. *Communication Research, 14,* 85–108.
Robichaud, D. (2001). Interaction as a text: A semiotic look at an organizing process. *American Journal of Semiotics, 17,* 141–161.
Roche, D. (1988). *Les républicains des lettres, gens de culture et lumières au XVIIIe siècle* [The members of the Republicans of letters, people of culture and enlightenment in the 18th century] (Vol. 18). Paris: Fayard.
Rousseau, J.-J. (1959). *Les confessions* [The confessions]. Paris: Gallimard Jeunesse. (Original work published 1763)
Rousseau, J.-J. (1990). *Essai sur l'origine des langues* [Essay on the origin of languages]. Paris: Folio Essai. (Original work published 1781)
Rousset, J. (1966). *Forme et signification* [Form and signification]. Paris: José Corti.
Sassen, S. (2002). *Global cities and diasporic networks: Microsites in global civil society.* Oxford, UK: Oxford University Press.
Schall, M. S. (1983). A communication-rules approach to organizational culture. *Administrative Science Quarterly, 28,* 557–581.
Schön, D. A. (1983). *The reflective practitioner: How professionals think in action.* New York: Basic Books.
Schunn, C., Crowley, K., & Okada., T. (2002). What makes collaborations across a distance succeed? The case of the cognitive science community. In P. Hinds & S. Kiesler (Eds.), *Distributed work: New research on working across distance using technology* (pp. 407–430). Cambridge, MA: MIT Press.
Searle, J. (2010). *Making the social world: The structure of human civilization.* Oxford, UK: Oxford University Press.
Seligman, M. (2009, May 24). One husband, two kids, three deployments. *New York Times,* A19. Retrieved from http://www.nytimes.com/2009/05/25/opinion/25seligman.html

Sellen, A. J., & Harper, R. H. R. (2002). *The myth of the paperless office.* Boston, MA: Massachusetts Institute of Technology.
Sellers, S. (2000). Virginia Woolf's diaries and letters. In S. Roe & S. Sellers (Eds.), *The Cambridge companion to Virginia Woolf* (pp. 109–126). Cambridge, UK: Cambridge University Press.
Shaw, G., Brown, R., & Bromiley, P. (1998, May). Strategic stories: How 3M is rewriting business planning. *Harvard Business Review, 76*(3), 46–50.
Short, J. A., Williams, E., & Christie, B. (1976). *The social psychology of telecommunications.* New York: John Wiley and Sons.
Siegel, J., Dubrovsky, V., Kiesler, S., & McGuire, T. W. (1986). Group processes in computer-mediated communication. *Organizational Behavior & Human Decision Processes, 37,* 157–187.
Smith, M. A. (1999). Invisible crowds in cyberspace. In M. A. Smith & P. Kollock (Eds.), *Communities in cyberspace* (pp. 195–219). New York: Routledge.
Smyth, J., & Pennebaker, J. W. (1999). Sharing one's story: Translating emotional experiences into words as a coping tool. In C. R. Snyder (Ed.), *Coping: The psychology of what works.* New York: Oxford University Press.
Sproull, L., & Faraj, S. (1995). Atheism, sex and databases: The net as social technology. In B. Kahin & J. Keller (Eds.), *Public access to the internet* (pp. 62–81). Cambridge, MA: MIT Press.
Sproull, L., & Kiesler, S. (1986). Reducing social context cues: Electronic mail in organizational communication. *Management Science, 32,* 1492–1512.
Sproull, L., & Kiesler, S. B. (1991). *Connections: New ways of working in the networked organization.* Cambridge, MA: MIT Press.
Stamps, J., & Lipnack, J. (2005, February). *Hubs in the diamond: The new science of organization networks* [White paper]. Retrieved from http://netage.com/pub/whpapers/Hubs-in-Diamond_NetAge-wp2.pdf
Stanton, A. L., & Danoff-Burg, S. (2002). Emotional expression, expressive writing, and cancer. In S. J. Lepore & J. M. Smyth (Eds.), *The writing cure* (pp. 31–51). Washington, DC: American Psychological Association.
Star, S. L., & Ruhleder, K. (1996). Steps toward an ecology of infrastructure: Design and access for large information spaces. *Information Systems Research, 7,* 111–134.
Stigliani, I., & Ravasi, D. (2009). Organizing thoughts and connecting brains: Material practices and the transition from individual to group-level prospective sensemaking. *Academy of Management Journal.*
Stoll, C. (1995). *Silicon snake oil: Second thoughts on information technology.* New York: Doubleday.
Suchman, L. A. (2007). *Human–machine reconfigurations: Plans and situated actions* (2nd ed.). Cambridge, UK: Cambridge University Press.
Tarrow, S. (1994). *Power in movement: Social movements, collective action and politics.* New York: Cambridge University Press.
Taylor, J. R., & Van Every, E. J. (2000). *The emergent organization. Communication as site and surface.* Hillsdale, NJ: Erlbaum.
Taylor, J. R., & Van Every, E. J. (2011). *The situated organization: Studies in the pragmatics of communication.* New York: Routledge.
Tidwell, L. C., & Walther, J. B. (2002). Computer-mediated communication effects on disclosure, impressions, and interpersonal evaluations: Getting to know one another a bit at a time. *Human Communication Research, 28,* 317–348.

Tönnies, F. (1967). Gemeinschaft and Gesellschaft. In C. Bell & H. Newby (Eds.), *The sociology of community* (pp. 7–12). London: Frank Cass and Co.

Trautmann Banks, J. (1990). *The selected letters of Virginia Woolf* [Introduction]. San Diego, CA: Harcourt Brace Jovanovich.

Tufte, E. R. (2006). *The cognitive style of PowerPoint: Pitching out corrupts within.* Cheshire, CT: Graphics Press.

Turkle, S. (1995). *Life on the screen: Identity in the age of the internet.* New York: Simon & Schuster.

Turkle, S. (2011). *Alone together: Why we expect more from technology and less from each other.* Philadelphia, PA: Basic Books.

Maanen, J. V. and Barley, S. R. (1985). Cultural Organization: Fragments of a Theory, in P. J. Frost, L. F. Moore, M. R. Louis, C. C. Lundberg & J. Martin (Eds.), *Organizational Culture* (pp. 31–53). Beverly Hills: Sage.

Vargas, J. G., & Torr, D. G. (1999). The Cartan-Einstein unification with teleparallelism and the discrepant measurements of Newton's constant G. *Foundations of Physics, 29*(2), 145–200.

Vincent, J. (2006). Emotional attachment and mobile phones. In S. Bertschi & P. Glotz (Eds.), *Thumb culture: The meaning of mobile phones for society* (pp. 117–122). New Brunswick, UK: Transcript Verlag.

Von Hippel, E., & Krogh, V. (2003). Open source software and the "private-collective" innovation model: Issues for organization science. *Organization Science, 14,* 209–223.

Vygotsky, L. (1962). *Thoughts and language.* Cambridge, MA: MIT Press.

Walsh, J. P., & Bayma, T. (1996). Computer networks and scientific work. *Social Study of Science, 26,* 661–703.

Walsh, J. P., & Ungson, G. R. (1991). Organizational memory. *Academy of Management Review, 16,* 57–91.

Walther, J. B. (1995). Relational aspects of computer-mediated communication: Experimental observations over time. *Organization Science, 6,* 186–203.

Walther, J. B. (1996). Computer-mediated communication: Impersonal, interpersonal, and hyperpersonal interaction. *Communication Research, 23,* 3–43.

Walther, J. B., & Burgoon, J. K. (1992). Relational communication in computer-mediated interaction: A meta-analysis of social and antisocial communication. *Communication Research, 21,* 460–487.

Wasko, M., & Faraj, S. (2000). It is what one does: Why people participate and help others in electronic communities of practice. *Journal of Strategic Information Systems, 9,* 155–173.

Weber, M. (1968). *Economy and society.* Berkeley: University of California Press.

Weeks, J., & Fayard, A-L. (2011). Blurring face-to-face and virtual encounters. *Harvard Business Review Blog Network.* Retrieved from http://blogs.hbr.org/cs/2011/07/blurring_face-to-face_and_virt.html

Weick, K. (1987). Theorizing about organizational communication. In F. Jablin, L. Putnam, K. Roberts, & L. Porter (Eds.), *Handbook of organizational communication* (pp. 97–122). Newbury Park, CA: Sage.

Weiner, M., & Mehrabian, A. (1968). *Language within language: Immediacy, a channel in verbal communication.* New York: Appleton-Century-Crofts.

Wenger, E., White, N., & Smith, J. D. (2009). *Digital habitats: Stewarding technology for communities.* Portland, OR: CPsquare.

White, H. C. (1992). *Identity and control: A structural theory of social action*. Princeton, NJ: Princeton University Press.
Wiesenfeld, B. M., Raghuram, S., & Garud, R. (1999). Communication patterns as determinants of organizational identification in a virtual organization. *Organization Science, 19*, 777–790.
Williams, A. (2009, June 21). Mind your Blackberry or mind your manner. *New York Times*, p. A1. Retrieved from http://www.nytimes.com/2009/06/22/us/22smartphones.html?_r=1&th&emc=th
Wilson, J. M., O'Leary, M. B., Metiu, A., & Jett, Q. R. (2008). Perceived proximity in virtual work: Explaining the paradox of far-but-close. *Organization Studies, 29*, 979–1002.
Wilson, J. M., Strauss, S. G., & McEvily, B. (2006). All in due time: The development of trust in computer-mediated and face-to-face teams. *Organizational Behavior and Human Decision Processes, 99*, 16–33.
Wolf, M. (2008). *Proust and the squid: The story and science of the reading brain*. New York: HarperCollins.
Yannis, G. (2008). Against the tyranny of PowerPoint: Technology-in-use and technology abuse. *Organizational Studies, 29*, 255–276.
Yates, J. (1989). *Control through communication: The rise of system in American management*. Baltimore, MD: Johns Hopkins University Press.
Yates, J., & Orlikowski, W. J. (1992). Genres of organizational communication: A structurational approach to studying communication and media. *The Academy of Management Review, 17*, 299–327.
Yates, J., & Orlikowski, W. J. (2007). The PowerPoint presentation and its corollaries: How genres shape communicative action in organizations. In M. Zachry & C. Thralls (Eds.), *The cultural turn: Communicative practices in workplaces and the professions* (pp. 67–92). Amityville, NY: Baywood.
Zander, U., & Kogut, B. (1995). Knowledge and the speed of the transfer and imitation of organizational capabilities: An empirical test. *Organization Science, 6*, 76–92.
Zhu, E. (1999). Hypermedia interface design: The effects of number of links and granularity of nodes. *Journal of Educational Multimedia and Hypermedia Archive, 8*, 331–358.

INDEX

A

Abelard 31, 32
Academia Parisiensis (French academy) 89
accountability building 66, 68–70
addressing reader mechanism: articulating emotions and 51–4; community building and 87; context, providing, and 66, 70–1; debating ideas and 75, 76–7; devfs-why not? thread and 128; e-mail and 117–18, 154–5; emotions expression and 21–2, 34–5, 134–5; empathy seeking and 47–9; knowledge development and 154–5; Mersenne and Descartes and 91; new theory articulation and 75, 77–8; online writing and 117–18; thought and 21–2; true dialogue engagement and 52, 55–7; trust building and 134–5; understanding emotions and 52, 54–5; of writing 21–2, 23
Albany Fort Standard of Trade example 41
Alexandria 61, 196
Alone Together (Turkle) 185
Anderson, Greg 195
Aristotle 199
ARPANET 143
articulating actionable concepts *see* concepts, articulating actionable
articulating emotions 51–4
articulating new theory 75, 77–8

B

Banks, Trautmann 56
Bazerman, C. 7
Bell, Vanessa 50
billet genre, Twitter and 174
Blaye, Adam 45
Bloomsbury Group 50, 82
Bone Marrow Donor challenge 151
book structure: historical perspective and 14, 15; mode emphasis in 15–16; overview 14–16; socio-material practices and 14–15
Born, Max 72, 131, 198
boundary blurring 114
Bowling Alone (Putnam) 164
brain 179, 199
Brenan, Gerald 54, 56
Broadbent, Stefana 188–9
building communities *see* community building

C

Calle, Sophie 135–6
cancer support groups 162
Carr, Nicholas 118, 164–5, 192
Cartan, Élie 74; Finslerian teleparallelism and 72–3
Cartan-Einstein theory: correspondence developing 71–8; debating ideas and 75, 76–7; generating ideas and 74–6; new theory articulation and 75, 77–8
Cassini, Giovanni Domenico 84
cat 136–8
censorship 93

challenges, OpenIDEO 145; Bone Marrow Donor challenge 151; Food Production and Consumption challenge 146–7, 150, 151; Maternal Health challenge 147; Sanitation Challenge 151; Social Business Challenge 150
Chayko, M. 174
Chinese Letters (Ying) 57
Christina of Sweden 62
Cicero 197
citizen science 13
"The Cognitive Style of PowerPoint" (Tufte) 157
Collins, R. 85
Columbia space shuttle 10, 157
communication 24; community building and 82–3; organizational 186–9; trends regarding 177; *see also* face-to-face communication; mode of communication; oral communication
communism 84, 197
community building: communication and 82–3; experimentation and 99–100; letters genre and 84–5; mechanisms of writing and 87; online: Chayko and 174; connecting through writing and 171–5; contradictory findings on 162–5; Facebook and 174; findings contradicting 163; findings favoring 163–4; HomeNet project and 163, 164; KM Forum and 165–71; OpenIDEO and 173–5; open source software and 172–3; overview about 161–2; professional development forums and 165; relationship strengthening and 169–71; self-referring and 166–7; shared history building and 168–9; shared identity and 162–3; social media and 174–5; overview about 81–2; as power of writing 27–8; scientific contributions and 100–1; supporting relationships and 101–4; *see also* du Châtelet, Émilie; Mersenne, Marin
company man identity 48
complaining 39, 41–5
concepts, articulating actionable: features and guidelines supporting 149–50; as major goal 149; OpenIDEO and 149–51; refining phase and 150–1
Condamine, Charles-Marie de la 103
Connecting: How We Form Social Bonds and Communities in the Internet Age (Chayko) 174
Control Through Communication (Yates) 24
Coser, L. 197
creative process: brain and 179, 199; of knowledge development 79–80; writing as 179–80, 199
criticizing 39, 41–5
crowdsourcing 13
Curie, Marie 197
cut and paste 155–6

D

dead signs 6–7, 33, 179
debating ideas 75, 76–7
deconstructionists 7
Descartes, René: Elisabeth of Bohemia and 62; knowledge development and 62, 196; mathematics and philosophy and 196; mechanisms of writing and 91; *Méditations* and 62, 92–3; Mersenne and 89–93; portrait of 92
developing knowledge *see* knowledge development
devfs-why not? thread: flame overview 126–7; mechanisms of writing and 128–9; negative emotions expressed in 127–8; rationality appeal in 128–30; true dialogue engagement in 130–1
dialectics 181–2, 195
dialogue: face-to-face communication and 32–3; true dialogue engagement and 52, 55–7, 130–1, 196
Dickinson, Violet 57

Digital Habitats: Stewarding Technology for Communities (Wenger, White, & Smith) 164
disinterestedness 84, 197
drawing 182–3
du Châtelet, Émilie: accomplishments of 94; Du Fay and 102–3; information and 101–4; *Institutions de Physique* and 94; Mairan and 100–1; Maupertuis and 97–8, 99, 100; mechanisms of writing and 99–100; portrait of 96; relationships and 101–4; science and 95–8, 100–1; scientific proof and 100–1; Voltaire and 95, 101, 103
Du Fay, Charles François de Cisternay 102–3, 198
du Pierry, Madame 94

E

East India Company 41–2, 49
Economy and Society (Weber) 24
Einstein, Albert 73; background about 71–2; Born and 72, 131, 198; correspondences of 72; Finslerian teleparallelism and 72–3; power law and 196; *see also* Cartan-Einstein theory
Elisabeth of Bohemia 62, 196
e-mail: addressing reader mechanism and 117–18, 154–5; cat 136–8; emotional state and 133; flaming and 131–2; genres of 115; letters and 115, 138–9; mechanisms of writing and 143–4; media and 113; message archiving and 143; objectifying mechanism and 116–17, 143–4, 153–4; personal 138–9; reflecting mechanism and 118–19, 155–6; specifying mechanism and 119, 156; statistics on 109–10
emotion, expressing: addressing reader and 21–2, 34–5, 134–5; articulating emotions and 51–4; criticizing and complaining 39, 41–5; dialogue and 32–3; empathy seeking and 47–9; epistolary novels and 57, 58; generality and 33; HBC and 36–49; health benefits of 35; justifying practices and 45–7; letter correspondence and 31–2; *marraines de guerre* and 32; objectifying mechanism and 20–1, 33–5; in organizational letters 36–49; overview about 31–2; in personal letters 49–59; as power of writing 26–7; reflecting mechanism and 22, 33–5; specifying mechanism and 22–3, 35–6; transforming to positive emotions 45–7; true dialogue engagement and 52, 55–7; trust building and 47–9; understanding emotions and 52, 54–5; Woolf's practices of 51–7; writing mechanisms and 32–5, 51–7
emotions online, expressing: addressing mechanism and 134–5; criticism and 127–8; debate about 122–4; devfs-why not? thread and 126–31; e-mail and 133; flaming and 131–2; hacker subculture and 131–2; Linux forum and 124–32; media richness theory and 122; objectifying mechanism and 132–4; overview about 121–2; positive emotion and 135–8; rationality appeals and 128–30; reflecting mechanism and 135–6; social information processing theory and 123–4; social media and 185–6; social presence theory and 122–3; software development communities and 125; specifying mechanism and 136–8; true dialogue and 130–1; trust building and 132–8
empathy seeking: company man identity and 48, 49; difficulties of everyday life and 47–8; distributed organizing and 48–9; trust building and 47–9
Encyclopedists 94, 197
epistolary novels 57, 58

experimentation 99–100
expressing emotions *see* emotion, expressing; emotions online, expressing
extinct languages 195

F

Facebook 189; community building and 174; statistics on 109
face-to-face communication: boundary blurring and 114; dialogue and 32–3; Republic of Letters and 83–4; as superior 5–6
feedback, giving 148–9
Finslerian teleparallelism 72–3
The First Cities and States: A History of the World (BBC program) 4
Fitzhugh, William H. 178
flaming: defined 125; devfs-why not? thread 126–31; e-mails 131–2; hacker subculture and 131–2; OpenIDEO and 148
FlavourCrusador 151
Food Production and Consumption challenge 146–7, 150, 151
forwarding 168–9
free and open source software development communities 125

G

Gaba, Rhodit 170–1
Galaxy Zoo 142
generating ideas *see* idea generation
giving feedback *see* feedback, giving
Golden, Catherine J. 10
Gopal (KM Forum founder) 166, 169, 170, 173
Greeks, alphabet of 18–19

H

hacker subculture 131–2
Hammurabi 4
Harrison, David K. 195
Havelock, Eric 17
HBC *see* Hudson's Bay Company
Heloise 31, 32
helping 170–1
Henderson, K. 183
history 4, 7; building, shared 168–9; invention of writing and 199; oral cultures and 168, 199; perspective of 14, 15
The History and Power of Writing (Martin) 20–1, 112–13
HomeNet project 163, 164
Hudson Bay map 37
Hudson's Bay Company (HBC): accountability building and 66, 68–70; Albany Fort Standard of Trade example 41; company man identity and 48, 49; context, providing, and 66, 70–1; criticizing and complaining and 39, 41–5; emotional expression practices and 38–9, 42; emotions expressed via letters of 36–49; empathy seeking and 47–9; justifying practices and 45–7; letter types produced by 36–8; Moose Factory letter 40; negative emotions expression in 39, 41–5; organizational memory development within 65–71; overview about 36; repository creation and 66, 67–8; sharing knowledge within 65–71; transforming to positive emotions and 45–7; trust building and 47–9; verbal *vs.* written order and 43; York Factory engravings 38, 39
Human Genome Project 143–4
hybrid interactional spaces 189–90

I

idea generation: objectifying mechanism and 74–6, 146–7; OpenIDEO and 146–8; reflecting mechanism and 74–6; specifying mechanism and 147–8
ideas, debating 75, 76–7
IDEO 144–5
innovation 191–2
Institutions de Physique (du Châtelet) 94, 100–1

intermediaries 88
Internet 184–5; *see also* community building, online; emotions online, expressing; knowledge development, online; online writing
"I Tweet, Therefore I Am" (Orenstein) 185–6

J

JAMA (clean-birth kit) 147
justifying practices 45–7

K

Kadushin, C. 85
kernel developers 125–6
KM Forum: analysis of 166; conclusions about 171; contributors 166; forwarding and 168–9; founding of 165; Gopal and 166, 169, 170, 173; linking and 168; overview about 165–6; performativity and 167; as professional development forum 165; quoting and 168–9; relationship strengthening and 169–71; self-referring and 166–7; shared history building and 168–9
knowledge development: addressing reader mechanism and 154–5; Cartan-Einstein theory and 71–8; contemporary organizations and 64; creative process of 79–80; Descartes and 62, 196; distributed organizations and 63–71; future research and 191–2; HBC and 65–71; mechanisms of writing and 62–3; online: concepts, articulating actionable, and 149–51; feedback and 148–9; idea generation and 146–8; mechanisms of writing and 153–6; objectifying mechanism and 153–4; in online contexts 152–6; at OpenIDEO 151–2; oral communication and 191–2; overview about 141–2; studies critical of 142; studies favoring 143; overview about 61–3; as power of writing 27; process-oriented approach and 65; reflecting mechanism and 155–6; role of writing in 63–5; science and 64, 71–8; sharing of knowledge and 63–4; specifying mechanism and 156; tacit knowledge and 65
knowledge management *see* KM Forum
Koenig, Samuel 100–1

L

language: extinct 195; oral cultures and 3, 195; as socially constructed 7
Lavoisier, Antoine-Laurent de 94
Lavoisier, Madame 94
leisure 191, 199
Lepaute, Hortense 94
letters 10; *Chinese Letters* 57; community building and 84–5; e-mails and 115, 138–9; emotion expressed in 31–2, 36–49, 49–59; epistolary novels and 57, 58; famous examples of 10, 31–2; genres of 84–5, 115; methodological approach and 28–9; mode of communication and 12; Moose Factory 40; personal e-mails and 138–9; private and public spheres and 85–8, 197; as proofs of scientific contributions 100–1; scientific communities and 84–5; as space for experimentation 99–100; supporting relationships via 101–4; wife and soldier 8–9; *see also* Cartan-Einstein theory; du Châtelet, Émilie; Hudson's Bay Company; Mersenne, Marin; Republic of Letters; Woolf, Virginia
Letters of Abelard and Heloise 31, 32
Letters of a Portuguese Nun 58
Life on the Screen (Turkle) 121
linking 168
LinuxChix 173

Linux forum: devfs-why not? flame in 126–31; emotions and 124–32; kernel developers 125–6; rationale for choosing 124–5; software development communities and 125
Linux operating system 172
Linux user groups 173
literacy perspective 17
Locker, Kitty 41
logic, oral cultures and 18

M

MacGregor, Neil 4
McLuhan, M. 192
Mairan (mathematician) 99, 100–1
marraines de guerre (war godmothers) 32, 199
Martin, Henri-Jean 20–1, 112–13
Massey, Doris 180
Maternal Health challenge 147
Maupertuis, Pierre-Louis 83, 97–8, 99, 100
mechanisms of writing: addressing reader 21–2, 23; changes in enactment of 113–14, 116; community building and 87; concepts, articulating actionable, and 149–51; criticizing and complaining and 39, 41–5; Descartes and 91; devfs-why not? thread and 128–9; du Châtelet and 99–100; e-mail and 143–4; emotions expression and 32–5, 51–7; empathy seeking and 47–9; enactment 113–14, 116, 158–9; feedback and 148–9; idea generation and 146–8; infrastructure changes and 113–14; justifying practices and 45–7; knowledge development and 62–3, 153–6; knowledge practices and 62–3, 66; Mersenne and 91; objectifying 20–1, 23; online writing and 113–14, 116, 143–4; overview about 20, 23–4, 109; reflecting 22, 23; Socrates's dialectics and 181–2; specifying 22–3; technology features and 114, 116; trust building and 47–9; using new media to enact 158–9; *see also specific mechanism*
media: beyond 192–3; addressing reader mechanism and 117–18, 154–5; boundary blurring and 114; Broadbent and 188–9; e-mail and 113; hybrid interactional spaces and 189–90; kinds of 12; knowledge development studies and 142–3; McLuhan and 192; mechanisms of writing enactment and 113–14, 116, 158–9; mode of communication and 11–13; objectifying mechanism and 116–17, 153–4; potential loss and 183–6; powers of writing and 114; public 86, 88; reflecting mechanism and 118–19, 155–6; richness scholars 6; richness theory 24–6, 122; social 154, 174–5, 185–6; social information processing theory and 123–4; social presence theory and 122–3; socio-material implications of 14–15, 111–13; specifying mechanism and 119, 156; speed of change in 113; trends regarding 177
Méditations (Descartes) 62, 92–3
memo 24–5
Mersenne, Marin: *Academia Parisiensis* initiated by 89; background about 88; censorship and 93; Descartes and 89–93; mechanisms of writing and 91; network of correspondences 88–91; Peiresc and 81, 89; volume of letters sent and received by 88–9, 90
Merton, R. 84, 197
Mesopotamia 4, 195
methodological approach: discourse analysis and 28, 29; empirical contexts of 29; letter writing and 28–9; phases of 29–30
mode of communication: book structure and 15–16; letters and 12;

media and 11–13; perspective 11; power of writing and 13, 20–1; relevance of 104–5; technology and 12; *see also* mechanisms of writing
Moose Factory letter 40
Mrs. Dalloway (Woolf) 50

N

naming a ship 199
negative emotions 39, 41–5, 127–8
Newell, Alan 61, 79–80
New Science 83–5, 197
new theory articulation 75, 77–8
Newton, Isaac 94
Noa 136–8
Nokia 145

O

objectifying mechanism: accountability building and 66, 68–70; articulating emotions and 51–4; community building and 87; debating ideas and 75, 76–7; Descartes and 91; devfs-why not? thread and 128–9; du Châtelet and 100–1; e-mail and 116–17, 143–4, 153–4; emotions expression and 20–1, 33–5, 132–4; idea generation and 74–6, 146–7; justifying practices and 45–7; knowledge development and 153–4; media and 116–17, 153–4; Mersenne and 91; new theory articulation and 75, 77–8; online writing and 116–17; repository creation and 66, 67–8; technology and 116–17, 153–4; true dialogue engagement and 52, 55–7; trust building and 132–4; understanding emotions and 52, 54–5; of writing 20–1, 23
Ong, W. J. 177
online writing: addressing reader mechanism and 117–18; boundary blurring and 114; devfs-why not? thread and 126–31; e-mail genres 115; flaming e-mails and 131–2; hacker subculture and 131–2; infrastructure and 113–14; Linux forum and 124–32; mechanisms of writing and 113–14, 116, 143–4; media richness theory and 122; objectifying mechanism and 116–17; overview about 109–11; permanence of 143–4; powers of writing and 114; reflecting mechanism and 118–19; second orality and 110; social information processing theory and 123–4; social presence theory and 122–3; specifying mechanism and 119; statistics on 109–10; technology features and 114, 116; trust building in 132–8; *see also* emotions online, expressing
opacity 195
OpenIDEO: challenges and 145, 146–7, 150, 151; collaborative creation and 146; community building and 173–5; concepts, articulating actionable, and 149–51; feedback and 148–9; flaming and 148; idea generation and 146–8; IDEO and 144–5; knowledge development at 151–2; project phases 145; social media and 154; team role 173–4
open innovation communities 13
open source software 13; community building and 172–3; development communities 125; social movement and 172; women and 173; *see also* Linux forum; Wikipedia
oral communication: knowledge development and innovation 191–2; Socrates's dialectics and 181–2; space and time and 180–1; *see also* face-to-face communication
oral cultures: history and 168, 199; language and 3, 195; logic and 18
Orenstein, P. 185–6
organizational communication 186–9

organizational memory development 65–71
organizational practice 111–12
organized skepticism 84, 197
Orlando (Woolf) 50
Ostrosky-Solis, Feggy 199
Oxfam 145

P

Passions of the Soul (Descartes) 62
Peiresc, Nicolas-Claude Fabri de 81, 89, 197
performativity 167, 199
Plato 6, 7, 20, 181–2, 195
Pliny 31
Pliny the Younger 197
positive emotion 135–8
Posting It: The Victorian Revolution in Letter Writing (Golden) 10
power law 196
PowerPoint slides: oral and written modes and 157–8; specifying mechanism and 156; "The Cognitive Style of PowerPoint" and 157
powers of writing: community building 27–8; contemporary debates about 25–6; emotions expression 26–7; knowledge development 27; Martin and 20–1, 112–13; mode of communication and 13, 20–1; online writing and 114; theories opposed to 24–6; *see also specific power*
praising 170
primitive notion 62
Principia (Descartes) 62
process-oriented approach 65
professional development forums 165
Putnam, Richard 164

Q

quoting 168–9

R

reading 19
recording 18
refining phase 150–1
reflecting mechanism: accountability building and 66, 68–70; community building and 87; cut and paste and 155–6; debating ideas and 75, 76–7; Descartes and 91; e-mail and 118–19, 155–6; emotions expression and 22, 33–5, 135–6; idea generation and 74–6; Internet and 184–5; knowledge development and 155–6; loss of 184–5; media and 118–19, 155–6; Mersenne and 91; new theory articulation and 75, 77–8; online writing and 118–19; technology and 118–19, 155–6; trust building and 135–6; understanding emotions and 52, 54–5; of writing 22, 23
relationships, supporting 101–4
relationship strengthening: activities of 169–70; helping as 170–1; inviting to contribute as 170; KM Forum and 169–71; praising as 170; thanking as 169, 170
repository creation 66, 67–8
Republic of Letters: conversation and 83–4; discursive practices of 83–4; experimentation and 99–100; intermediaries and 88; letters genre and 84–5; liminal status of letters and 85–8; overview about 82, 83; private and public spheres and 85–8; public media and 86, 88; supporting relationships and 101–4; women and 93–4, 103–4; writing practices used within 87; *see also* du Châtelet, Émilie; Mersenne, Marin
research: future 186–92; hybrid interactional spaces and 189–90; knowledge development and innovation 191–2; organizational communication future 186–9
A Room of One's Own (Woolf) 50
Rousseau, Jean-Jacques 32, 33
Royal Library of Alexandria 61, 196

S

salon 83–4
Sanitation Challenge 151
science: citizen 13; du Châtelet and 95–8, 100–1; knowledge development and 64, 71–8; New Science and 83–5, 197; scientific communities and 84–5; scientific contributions proof and 100–1
scriptura continua (writing with no separation between words) 184
second orality 110
self-referring: KM Forum and 166–7; as performative 167, 199; practice enactment 167
The Shallows (Carr) 118, 164–5
shared history building: discursive practices of 168; KM Forum and 168–9; writing and 168, 199
shared identity 162–3
Shaw, Cliff 61, 79–80
ship naming 199
Simon, Herbert 61, 79–80
skepticism, organized 84, 197
Skype 8–9
smart phones 10–11
Smith, Ethel 52–3, 54
Smith, J. D. 164
Smith, M. A. 163
Social Business Challenge 150
Social Change in a Box 151
social information processing theory 123–4
social media: community building and 174–5; connecting and 185–6; OpenIDEO and 154
social presence theory 25, 122–3
socio-material practices 14–15, 111–13
Socrates 181–2
soldier and wife letters 8–9
sophism 181–2, 199
Sophists 7, 181–2
space and time 180–1
specifying mechanism: accountability building and 66, 68–70; articulating emotions and 51–4; community building and 87; context, providing, and 66, 70–1; Descartes and 91; devfs-why not? thread and 128; e-mail and 119, 156; emotions expression and 22–3, 35–6, 136–8; empathy seeking and 47–9; idea generation and 147–8; knowledge development and 156; media and 119, 156; Mersenne and 91; online writing and 119; PowerPoint slides and 156; repository creation and 66, 67–8; technology and 119, 156; trust building and 136–8; of writing 22–3
Stephen, Leslie 50
stream of consciousness technique 50; *see also* Woolf, Virginia
strengthening relationships *see* relationship strengthening
structuralists 7

T

tablets 4, 195
tacit knowledge 65
Take Care of Yourself (Calle) 135–6
technology: addressing reader mechanism and 117–18, 154–5; boundary blurring and 114; feature changes 114, 116; knowledge development studies and 142–3; mechanisms of writing enactment and 113–14, 116; media richness theory and 122; mode of communication and 12; objectifying mechanism and 116–17, 153–4; potential loss and 183–6; powers of writing and 114; reflecting mechanism and 118–19, 155–6; Skype and 8–9; smart phones and 10–11; social information processing theory and 123–4; social presence theory and 122–3; socio-material implications of 14–15; space and time and 180–1; specifying mechanism and 119, 156; speed of change in 113; trends regarding 177

text messages, statistics on 6
thanking 169, 170
thought: abstract 18–19; addressing reader and 21–2; objectifying 20–1; reflecting and 22; specifying and 22–3; time for 19
time and space 180–1
Torvalds, Linus 172
To the Lighthouse (Woolf) 50
Trajan 31
true dialogue engagement 52, 55–7, 130–1, 196
trust building: addressing mechanism and 134–5; empathy seeking and 47–9; objectifying mechanism and 132–4; in online writing 132–8; reflecting mechanism and 135–6; specifying mechanism and 136–8
Tufte, E. R. 10, 157, 158
Turkle, Sherry 121, 185, 190
Twitter 185–6; billet genre and 174; statistics on 109

U

understanding emotions 52, 54–5
Understanding Media: The Extensions of Man (McLuhan) 192
Unilever 145
universalism 84, 197
Usenet groups 163

V

Virtual Vietnam 13, 162, 164
visual practices 182–3
Voltaire, François-Marie Arouet de 94, 95, 97, 101, 103

W

Walther, J. B. 123–4
Water and Sanitation for the Urban Poor (WSUP) 145
Weber, M. 24
WELL 82, 164
Wenger, E. 164
White, N. 164
wife and soldier letters 8–9
Wikipedia 142, 172
Wolf, Maryanne 19, 179, 193, 199
women: open source software and 173; Republic of Letters and 93–4, 103–4
Woolf, Virginia: articulating emotions and 51–4; background about 50; Bloomsbury Group and 82; correspondences of 50–1; emotional expression practices and 51–7; portrait of 51; stream of consciousness technique of 50; true dialogue engagement and 52, 55–7; understanding emotions and 52, 54–5
writing: connecting through 171–5; contradictory imaginations about 5–11; as creative process 179–80, 199; criticisms of, historical 7; as dead signs 6–7; Fitzhugh on 178; for granted assumptions about 9; overview about 3–4; power of 13, 20–1; powers of, three 24–8; process of 179–83; role of 4–11; shared history building and 168, 199; socio-material practices of 111–13; space and time and 180–1; speaking and 5–11; technology and 8–9; *see also* mechanisms of writing; online writing; *specific subject*
WSUP *see* Water and Sanitation for the Urban Poor

Y

Yates, J. 24
Ying Chen 57
York Factory engravings 38, 39